NUMBER FOUR:

The Texas Engineering Experiment Station Monograph Series

ANDREW R. MCFARLAND, *Editor-in-chief*

Texas A&M University

Converter Circuits for Superconductive Magnetic Energy Storage

Converter Circuits for Superconductive Magnetic Energy Storage

By Mehrdad Ehsani
and Robert L. Kustom

PUBLISHED FOR THE
TEXAS ENGINEERING EXPERIMENT STATION,
THE TEXAS A&M UNIVERSITY SYSTEM,
BY
TEXAS A&M UNIVERSITY PRESS
COLLEGE STATION

Manufactured in the United States of America
FIRST EDITION

The paper used in this book meets the minimum requirements of the American National Standard for Permanence of Paper for Printed Library Materials, Z39.48-1984. Binding materials have been chosen for durability.

Library of Congress Cataloging-in-Publication Data

Ehsani, Mehrdad.
Converter circuits for superconductive magnetic energy storage.

(The Texas Engineering Experiment Station monograph series; no. 4)
Includes bibliographies and index.
1. Magnetic energy storage. 2. Superconducting magnets. 3. Switching circuits. I. Kustom, Robert L. II. Title. III. Series.
TK2985.E37 1988 621.34 85-40055
ISBN 0-89096-257-X (alk. paper)

To our wives,
Millie and Dolores

Contents

Converter Circuits for Superconductive Magnetic Energy Storage

CHAPTER 1

Introduction

Superconductive magnet coils are increasingly being used for energy storage and generation of high magnetic fields in research and industry. Superconductive magnets are not only capable of producing high fields (10 T or more), but are also energy efficient due to their very low losses. Some examples of superconductive coils for generation of high magnetic fields are the poloidal field coils of the Tokamak power reactor and the accelerator magnets in the large super colliders. Examples of superconductive coils used for energy storage include the Wisconsin superconductive energy storage coil, which is being developed for utility power system diurnal energy storage, and the experimental power system stabilizer coil of Los Alamos National Laboratory, which has undergone system tests.

The unique feature of the superconductive coil that makes it attractive for production of a high magnetic field and bulk energy storage is its zero resistance to current. High electrical currents can be concentrated in small coil windings without excessive difficulty with the heat removal caused by I^2R losses. The superconductive coil is a high-Q inductor capable of storing large amounts of energy. To supply this energy efficiently, the power supply should also have low losses and be capable of reversible power control. The need for efficient reversible power control is particularly great for repetitively energized magnets. Solid-state-switching power supplies have been utilized in recent years for this purpose. The low conduction losses, efficient switching, and increasingly higher power ratings of solid-state power switches make these supplies favorable for superconductive magnet applications in the future.

This book describes and analyzes some of the recent work on the development of switching power converters for superconductive magnets. The authors have been involved with most of these projects, along with their colleagues at Argonne National Laboratory, the University of Wisconsin Applied Superconductivity Center, and Texas

A&M University. However, the works of other researchers have also been included in an attempt at completeness.

1.1 Magnetic Energy Storage

The converter networks described in this book are applied in two areas of superconductive magnetic energy storage (SMES).

One area of application is for very large, repetitively pulsed powers, hundreds to thousands of megawatts, into and out of an inductive load that requires fairly precise, active control of the current in the load. The average power consumed during a cycle is generally a small fraction of the total power during the pulse. The traditional means of supplying large pulsed power to inductive loads is with a motor-generated-flywheel (MGF) set. The motor is connected to the inductive load through a converter network. The energy to charge the inductive load is extracted from the rotating mass of the MGF. At the end of the operating period, energy from the load is returned to the MGF by driving the synchronous generator as a motor. The overall efficiency of an MGF is relatively low, and operation of the inductive load at full current is usually limited to less than a minute because of windage, friction, excitation, and other conduction losses.

The large inductive loads of the future are increasingly likely to be superconductive, and for these applications SMES has a number of advantages over an MGF as an energy buffer. SMES is far more efficient and will allow operation at full current and low power for long periods of time. SMES does not require constant supervision while in operation because it has no moving parts and for the same reason requires less maintenance. Also, SMES does not require a heavily braced floor, as is the case with an MGF.

The second area of application is for load leveling over a twenty-four-hour period. The only really major diurnal energy system currently in use is pumped hydroelectric power. Pumped hydroelectric is economical in areas where there is an adequate water supply and a suitable terrain for building the reservoir. The efficiency of returning the stored energy to the utility is usually between 60% and 75%. The losses are due to excitation of the pumps, motors, and generators; conduction losses in the leads; and evaporation.

Superconductive magnetic storage has the potential of being far more efficient, nearly 98%, and can be more universally sited. The converter networks for these applications must be able to reversibly and efficiently transfer the energy into and out of the SMES coil to make full use of the superconducting coil's low loss characteristic.

1.2 Superconductive Coils as Energy Buffers

With respect to the design of the converter network, a superconductive coil is treated as a virtually lossless inductor. Indeed, the electrical circuit losses of a superconductive coil are really negligible.

Modern superconducting wire is drawn with thousands of strands of superconductor imbedded inside a base material of copper or aluminum. The base material is known as the stabilizer and assists in maintaining superconductivity when the conductor experiences a sporadic heat pulse. The most common superconductors are the alloys niobium titanium, NbTi, and niobium tin, Nb_2Sn_3. The wire is wound in a coil and cooled to a temperature between 1.5 and 4.5 deg K. The coil is either immersed in liquid helium or the liquid is pumped through the conductor. Either way, the coil and helium are surrounded by a vacuum jacket, radiation shields, usually a liquid nitrogen shield, and another vacuum jacket.

There are no conduction losses in the wire. There are joints in the conductor which provide some electrical losses, and there are some eddy current losses in the stabilizer due to the time-varying magnetic field. However, these losses are virtually nonexistent with respect to the normal conducting losses of the room-temperature leads and the thyristor losses in the converter. They do, however, cause heating of the helium coolant; that heat must be extracted by the refrigeration system. The power required for refrigeration is a major part of the system losses, whereas the direct electrical losses in the coil are not.

1.3 Switching Elements for Superconductive Magnet Power Supplies

Switching power converters have an ideal efficiency of nearly 100%. Therefore, they are compatible with the lossless superconductive magnets. These converters consist of switches, capacitors, and inductors. The practical realization of these components introduces losses inherent with their constituent materials. Silicon-based power switches presently offer the most efficient switching element. How-

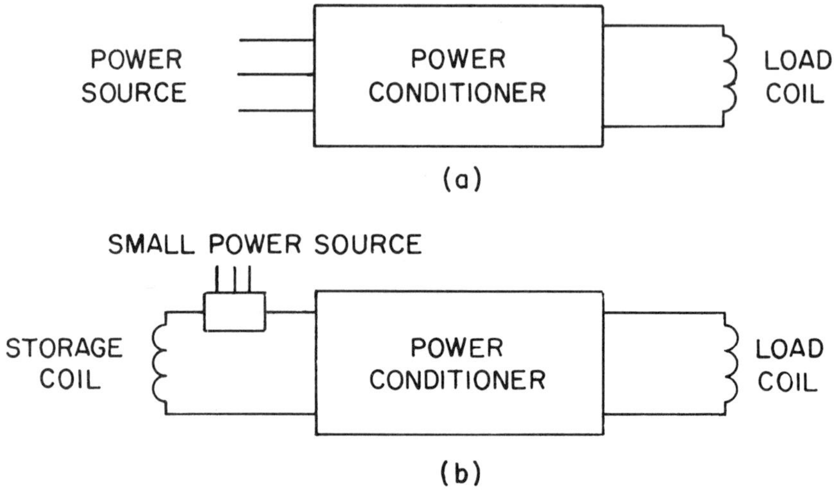

FIG. 1.1. Two basic power supply arrangements for inductive loads: (a) power conditioner supplies the variable power directly from the source; (b) variable power is supplied by exchanging energy between the storage and load coils. A small supply compensates the system for the losses.

ever, several structures of silicon-based switches exist, with various characteristics. Two broad structural groups are the three-layer devices, which include power transistors and power MOSFETs, and four-layer devices, which include thyristors. For the most efficient realization of a switching element the suitable choice of the type of solid-state switch must be based on the converter's operating characteristics.

Converter circuits for superconductive magnets are usually high-current systems. Furthermore, the required switching frequencies in these converters are at the lower end of the spectrum offered by the various types of solid-state switches. The most efficient realization of a high-current, low-frequency switch is the four-layer device: the silicon-controlled rectifier (SCR). The SCR presently has the highest current ratings and the lowest conduction losses of any solid-state switch and is expected to maintain this superiority into the foreseeable future. However, this device is not capable of turning off internally. For converter circuits in which interval turn-off is desirable, the gate-turn-off thyristor (GTO) is more suitable. The GTO, which is another four-layer device, is not as efficient in conduction as the SCR, but the GTO may achieve better overall efficiency if forced periodic turn-off is required.

The following sections give an introductory view of the SCR and the GTO. More comprehensive treatment of these devices appears in the references listed at the end of this chapter.

1.4 The SCR

Thyristors are p-n-p-n semiconductor devices which use regenerative feedback to produce bistable switching action. The most commonly used thyristor is the SCR.

The SCR is a three-terminal solid state switch that can be used in high-power circuits. The graphical symbol for an SCR device is shown in Figure 1.2, and a representation of the static anode-cathode characteristics is shown in Figure 1.3.

The main current-carrying connections are between the anode and the cathode. The gate carries only a very low level of current (0.1 A to 1.0 A). When voltage is applied in the reverse direction, the SCR passes a very low level of leakage current as long as the reverse voltage is below the reverse breakdown value. When voltage is applied in the

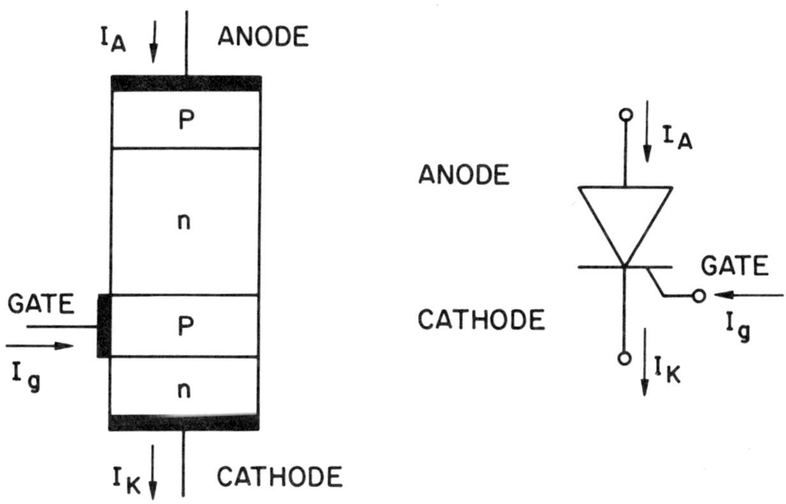

FIG. 1.2. Internal structure and schematic symbols for an SCR device.

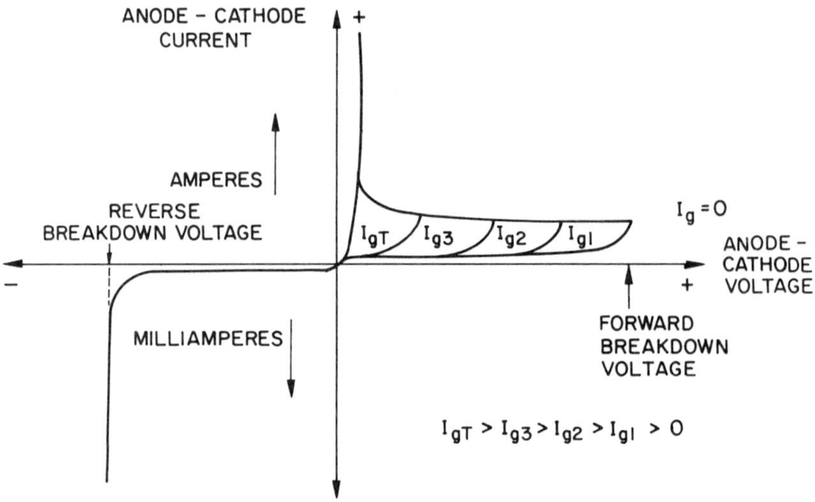

FIG. 1.3. Representation of static anode-cathode characteristics of the SCR.

forward direction, only a small leakage current flows until some critical value of voltage is reached. Then a high-current, very-low-impedance state is established with an extremely low ($\simeq$ 1.5 V) potential drop. The critical value of voltage is dependent on the gate current. In practical application, the only two levels of gate current that are used are zero and the saturated state, I_{gT}. Thus, as long as the operating voltage is below the forward breakdown voltage, the device carries little or no forward current until the gate is "triggered" by a current of I_{gT}.

Once the SCR is triggered into the conducting state, the gate loses control, and removal of the gate current does not influence the anode current, provided that the anode current exceeds a minimum current called the "latching current." In quasistatic operation, the forward block state is re-established by reducing the anode current to a relatively low level, almost zero, for a sufficiently long time for the carriers in the barrier region to recombine.

1.4.1 Dynamic Switching of SCRs

When a current pulse is delivered to the gate of the SCR, it does not immediately go into a conducting state. For a period of time, called the "delay time," t_d, the SCR continues to block current and hold the

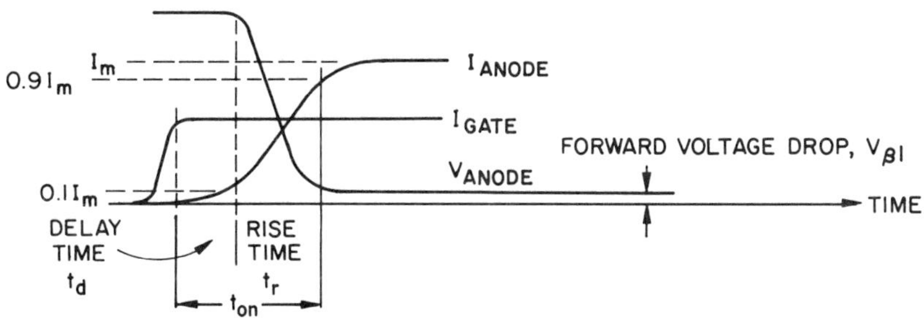

FIG. 1.4. Dynamic turn-on of the SCR.

anode-cathode voltage at the blocking value. Afterwards, the impedance starts to decrease over a period of time until the fully conducting state is reached. The time it takes for the current to rise from 10% to 90% of its peak value is called the rise time, t_r. The sum of the delay time and the rise time is called the turn-on time, t_{on}, depicted in Figure 1.4.

The shortest possible turn-on time requires a gate that rises to full current in 0.1 to 1.0 μs for an amplitude that is three to five times greater than the minimum gate pulse. The total turn-on times can be between 5 and 10 μs for the more rapid turn-on devices on the market. The SCR device has a practical limit for the maximum di/dt that can be tolerated without causing overheating. This limit results from the fact that an initial charge column is created along a narrow region of the barrier and requires time to transversely diffuse over the bulk of the barrier region.

The blocking condition of an SCR is not restored instantly. The current in the SCR must be reduced to zero, which will require discharging the lead inductance. The zero-current condition must be maintained for a sufficiently long time for the free charges in the barrier region to diffuse and/or recombine. The shortest time for turn-off is achieved by applying a reverse voltage to the SCR. In the first instance, after the lead inductance is discharged, the reverse voltage drives current in the negative direction through the SCR. The reverse current helps sweep the charges out of the barrier. After some time of reverse current flow, the SCR voltage starts to reverse, and this current will start to decrease towards zero. The reverse recovery time is defined as the period from when the anode current crosses zero in the

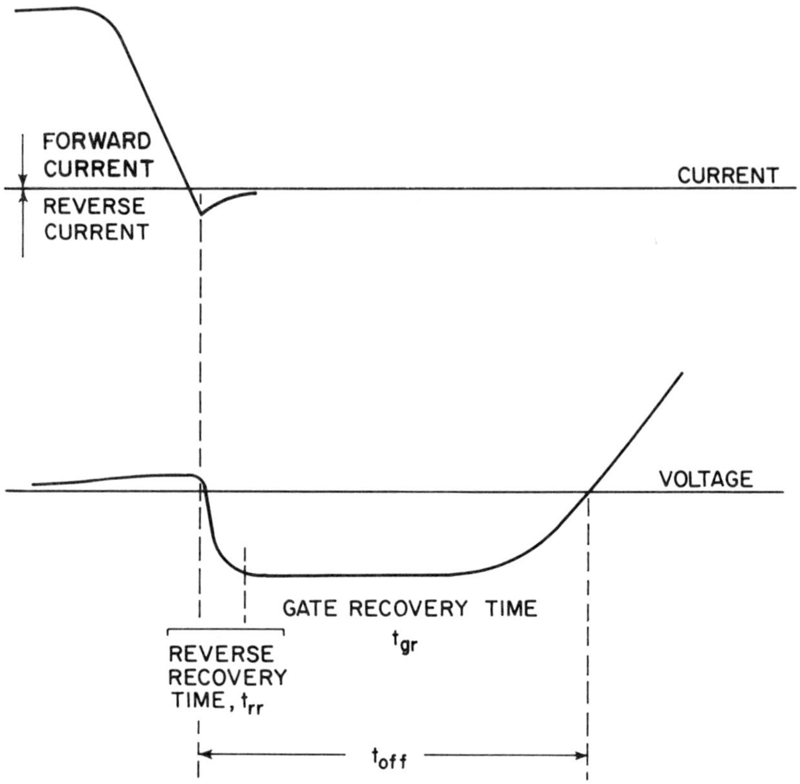

FIG. 1.5. Dynamic turn-off of the SCR.

reverse direction until it reaches zero again. Voltage still cannot be applied in the forward direction because enough free charges exist in the barrier to start forward conduction over. Some time is required for most of these charges to recombine. This time is called the "gate recovery time." The total turn-off time is the sum of reverse recovery time and the gate recovery time. This is shown in Figure 1.5.

The rate of rise of voltage, *dv/dt*, on an SCR must be below a specified minimum value. This is required to avoid spurious triggering that can be induced through capacitively injecting charges into the barrier region.

Since an SCR has a *di/dt* and a *dv/dt* limit, it is common to add a small series inductor into the lead and a small capacitor across the device to control these parameters. In addition, a small resistor is put in series with the capacitor to prevent a large current surge when the device is turned on. This latter circuit is known as a "snubber."

1.4.2 Parallel and Series Connections of SCRs

Parallel connections of SCRs are possible if one wants to increase the current rating of the switch beyond a single SCR device rating, but special consideration has to be given to balancing the currents between the individual devices because the drops in forward conducting voltage can vary significantly. Two common techniques are (1) to insert a small resistor in series with each SCR, or (2) to add a balancing reactor. The latter technique is usually preferred because it leads to considerably less power loss in the circuit. Figure 1.6 shows two SCRs with balancing reactors. If the forward voltage drops of S_1 and S_2 are different by the amount ΔV_β, the currents would be different by the amount $\Delta V_\beta / R$. The value of R would presumably be very small, leading to large imbalance without the balancing reactor. The currents would divide nearly equally if the balancing reactor provides a compensating voltage of ΔV_β with the right polarity. Thus, the reactor must provide

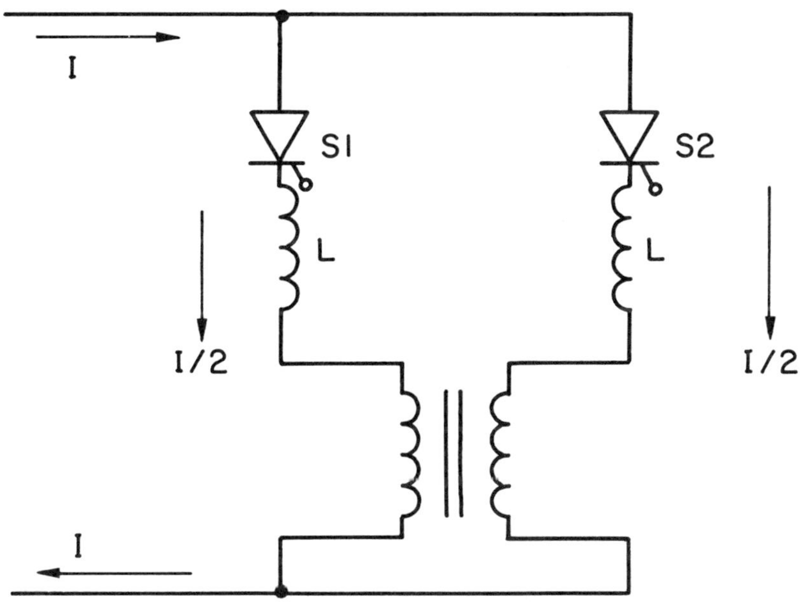

FIG. 1.6. Two SCRs in parallel with a balancing reactor.

$$\Delta V_\beta = Nd\phi/dt, \tag{1.1}$$

or

$$N\Delta\phi = \Delta V_\beta dt. \tag{1.2}$$

Thus, if the reactor is large enough to support $\Delta V_\beta dt$ volt-seconds of flux before saturation, the two SCR currents remain balanced.

A group of SCRs can be connected in series, provided impedance networks are added in parallel with each device to absorb voltage imbalances created by jitter in turn-on and turn-off times between individual devices. Figure 1.7 shows a group of SCRs connected in series. The capacitors used in the snubber networks must be large enough so that

$$V_{\max} > \frac{I\Delta t_{\text{off}}}{C}, \frac{I\Delta t_{\text{on}}}{C}, \tag{1.3}$$

where $V_{\max}$ is the maximum rating of an individual SCR. The resistor connected in parallel to the snubber network is needed if the switch is used in a circuit with a long-term direct current.

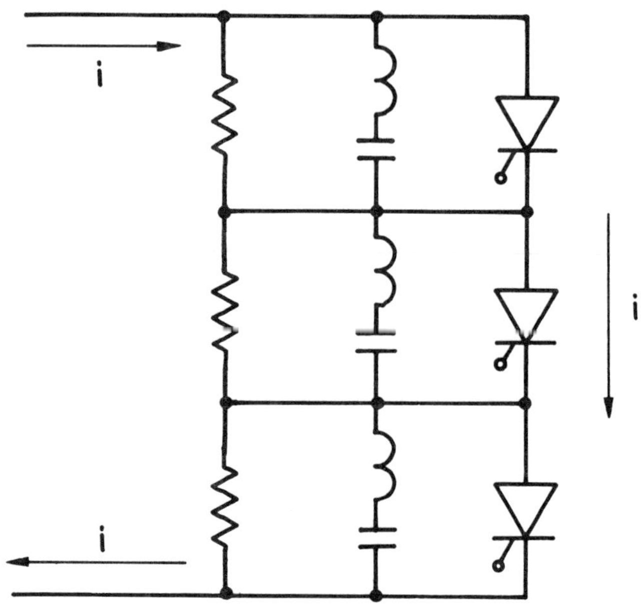

FIG. 1.7. Series connection of SCRs.

1.4.3 Commutation of the SCR

Since the SCR has no means of interrupting a current flowing in it, the process whereby it reverts to the nonconducting state is of fundamental importance in the application of the device. In some circuits the SCR current is diverted to an alternate path by the influence of the voltages operating in the circuit. SCR conduction ceases naturally as a result of this phenomenon, known as natural commutation. The uniform switching converters, introduced in chapter 4, provide such natural SCR commutations.

If natural commutation does not occur, an additional circuit is required to divert the current from the SCR and to allow it sufficient recovery time. This process is referred to as forced or artificial commutation. A simple conceptualization of the circuit for forced commutation is shown in Figure 1.8(a). The normal current-carrying element is thyristor T_1. Current I flows from the source on the left to the load on the right. To force commutation in T_1, the second thyristor, T_2, is gated into conduction. This causes the precharged capacitor, C, to be discharged around the local loop comprising the two thyristors and the loop inductor, L. The direction of this current is such as to oppose the existing current in T_1 and drive it to zero. Alternatively, the process can be thought of as one of transferring, or commutating, the load current from its path through T_1 into the capacitor.

The current i_1 will be sinusoidal in form and of a peak value $V_C(0)\sqrt{(C/L_1)}$, where $V_C(0)$ is the precharge voltage on the capacitor. This peak current must exceed the instantaneous value of the current I if T_1 interruption is to be effected. The load current actually continues to flow, for it has not been truly interrupted but rather commutated into the capacitor C. However, C will rapidly charge and establish a voltage opposing the load current, thereby driving it to zero. What we observe is an oscillation between the source inductor L (augmented by L_1), the capacitor C, and the load, which is interrupted by T_2 as soon as I attempts to reverse.

The design values of the forced commutation circuit components are based on the following considerations: The commutation loop inductor, L_1, must be great enough to assure that di_1/dt will not exceed the safe limit for the thyristor T_2. Referring to Figure 1.8(b), note that T_1 is subjected to an inverse voltage immediately after blocking at point A. The precharge commutation voltage $V_C(0)$ should be such that this inverse voltage is within the capability of the device. The inverse

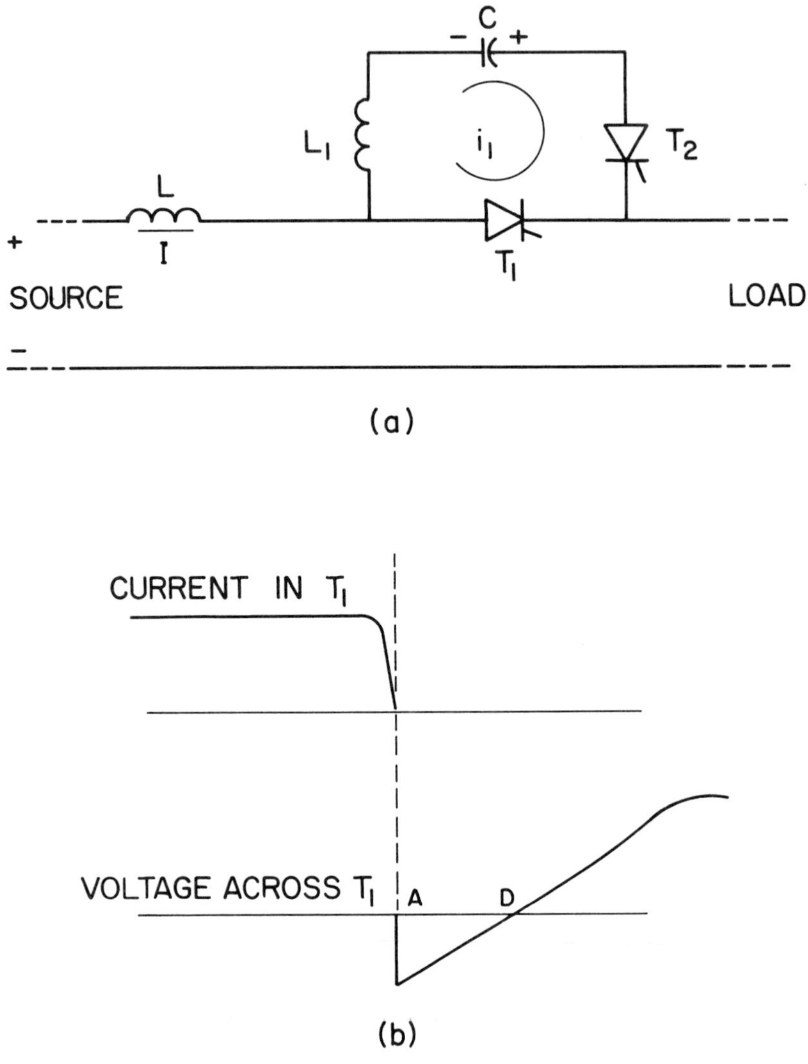

FIG. 1.8. Forced commutation of the SCR.

voltage persists across T_1 for a period *AD* (Figure 1.8[b]); thereafter it become positive. This time, *AD*, must be longer than the turn-off time of T_1. The design value of C is based primarily on this turn-off time.

Practical design of commutation circuits requires attention to several additional theoretical and practical details which will not be discussed here. However, important commutation considerations, particular to the circuits of this book, will be examined as these circuits are introduced.

1.5 The GTO

The gate of an SCR has no further control once the device has been triggered on; i.e., no gate current is required to maintain conduction in the SCR. However, it is possible to turn off certain types of thyristors by the application of a reverse gate current. This gate turn-off capability is very desirable because it makes the control of current for both its on time and off time possible without the use of elaborate commutation circuitry. In principle, all thyristor devices can be turned off by a reverse gate bias. However, thyristors designed specifically for this application are known as gate-turn-off thyristors (GTO). The schematic symbol of the GTO is shown in Figure 1.9.

One of the most important characteristics of the GTO is its maximum interruptable anode current. This parameter depends on the snubber circuit which is connected in parallel with the device, the gate drive circuit, and the device's internal structure. Interruptable anode currents of up to 600 A have been possible with GTOs for several years. A GTO with interruptable anode current of 2,000 A has recently been announced.

Another important GTO characteristic is its turn-off current gain, defined as $(I_A/I_{G,\,peak})$, where I_A is the on-state anode current and I_G is

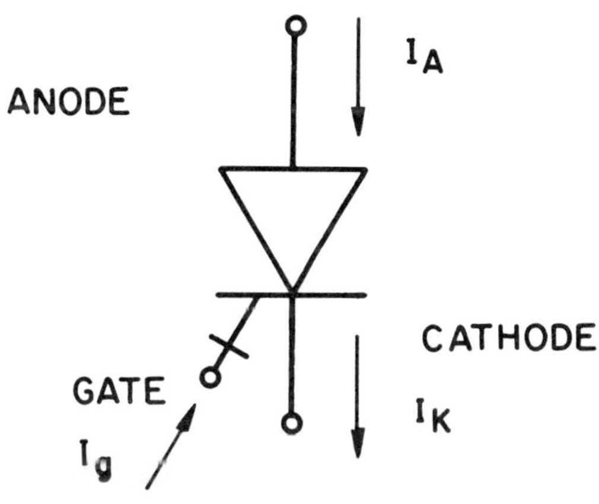

FIG. 1.9. Schematic symbol of the GTO.

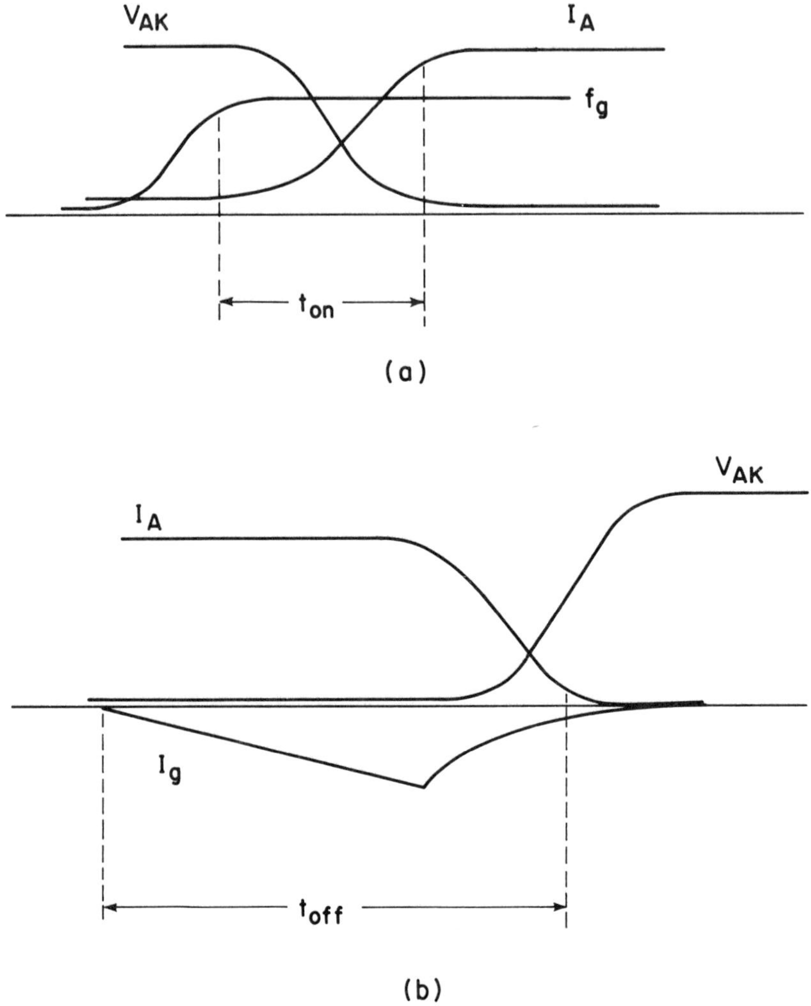

FIG. 1.10. Dynamic GTO switching waveforms: (a) turn-on; (b) turn-off.

the reverse gate current during the turn-off operation. This dynamic parameter depends on the rate of rise of gate current (dI_g/dt) and can be from 4 to 10. Therefore, the gate circuit must be capable of a substantial peak current for a short time. This is in contrast to the gate turn-on current for the SCR and GTO, which is just a few amperes for a short time. Typical GTO turn-on and turn-off waveforms are shown in Figure 1.10. Turn-on time of GTOs is around 10 μs and their turn-off time is around 12 μs. A severe recovery voltage spike can appear across the anode and cathode of the GTO during turn-off because of the trapped energy in the stray inductance of the power circuit. In order to minimize this spike, close attention is paid to the design of the GTO snubber circuit.

Recently, very large GTO devices have been announced (4.5 kV, 2 kA). These devices, which use anode-shorted emitter construction, provide high forward open voltage withstanding and low conduction voltage capabilities. However, these characteristics are achieved by sacrificing the reverse blocking voltage capability. Therefore, either fast antiparallel diodes are used with these GTOs or reverse voltage protection is provided by series diodes.

Direct paralleling of some GTOs can be realized by screening the on-state voltage and matching the switching times by coupling gate current. Series connection can be obtained by matching the switching times and forcing voltage sharing by paralleled resistors.

Other parameters of the GTOs, not discussed in this section, are substantially similar to those of the SCRs.

References

Argonne National Laboratory. *Tokamak Experimental Power Reactor Conceptual Design.* Report ANL/CTR-76-3. Argonne, Ill.: Argonne National Laboratory, 1976.

Blicher, A. *Thyristor Physics.* New York: Springer-Verlag, 1976.

Boom, R. W.; H. A. Peterson; et al. *Wisconsin Superconductive Energy Storage Project Report.* Vol. I. Madison: University of Wisconsin, 1974.

Ghandhi, S. K. *Semiconductor Power Devices.* New York: John Wiley and Sons, 1977.

Hassenzahl, W. V. *Electrochemical, Electrical, and Magnetic Storage of Energy.* Stroudsburg, Pa.: Hutchinson Ross Publishing Company, 1981.

Hitachi, Ltd. *Hitachi GTO Thyristor Catalog No. EC-E904P.* Tokyo, 1983.

Kustom, R. L., et al. *A Revised Design for the Tokamak Experimental Power Reactor.* Fusion Power Program, Report ANL/FPP/TM-77. Argonne, Ill.: Argonne National Laboratory, 1977.

Laquer, H. L.; F. L. Ribe; and D. M. Wilson. "Energy Storage and Switching with Superconductors as a Power Source for Magnetic Fields." In *Pulsed Thermonuclear Experiments and Reactors.* Boston, Mass.: Intersociety Energy Conversion Engineering Conference, 1971.

Mills, F. E. "The Fermilab Cryogenic Energy Storage System," *IEEE Transactions on Magnetics,* MAG-11, no. 2 (March 1975):489.

Sitting, R., and P. Roggwiller. *Semiconductor Devices for Power Conditioning*. New York and London: Plenum Press, 1982.

Stacey, W. M., Jr., et al. "Tokamak Experimental Power Reactors." In *Proceedings of the ASME Nuclear Engineering Division Annual Winter Meeting*. New York, December 5, 1976.

CHAPTER 2

Nonuniform Switching Circuits

The energy transfer converters proposed in this book may be classified into two distinct switching groups. Those discussed in this and the next chapter require continuous change of switching time intervals, throughout the energy transfer, to produce desired outputs. This group is named nonuniformly switching converters. The second group of converters, discussed in the later chapters, operate with a constant switching time interval throughout the energy transfer. These constant intervals are only altered during short switching transients caused by the controller. This group is named uniformly switching converters.

The simplest converter switching topologies belong to the nonuniformly switching group. As will be seen shortly, these converters can be synthesized with one or more switching elements. A switching element is defined in this book as a device which is capable of conducting current in one direction when in the *on* state. When the switching element is in the *off* state, it can withstand anode to cathode voltages of both polarities. Several actual switching devices may have to be connected in series and parallel to realize a switching element which has the required voltage and current capability. Note that no mention is made of the turn-off capability of a switching element in its definition. A switching element may have self-turn-off capability, such as that of a GTO, or external turn-off provisions may be necessary, as for an SCR. We will represent a switching element as shown in Figure 2.1(a). A switching element, enclosed in a circle shown in Figure 2.1(b), symbolizes forced turn-off capability either by the switch or by external means not shown in the diagram.

2.1 One-Switch Circuits

The simplest energy transfer circuit topologies are those which use only one switching element. Other nonswitch elements in the circuit will serve as a current buffer between the two magnets

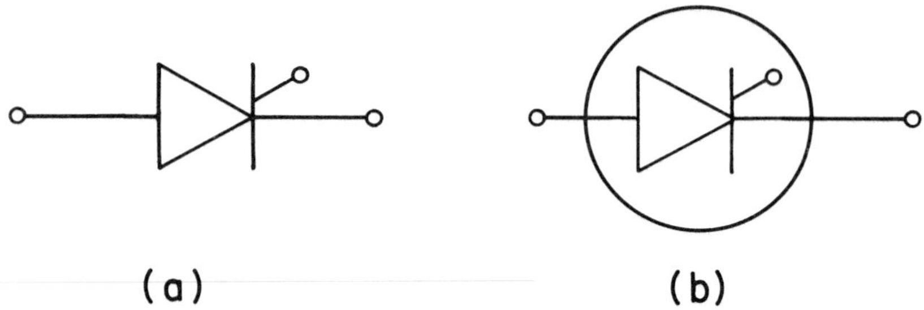

FIG. 2.1. Symbols for a switching element (a) with no forced turn-off capability and (b) with unspecified forced turn-off capability.

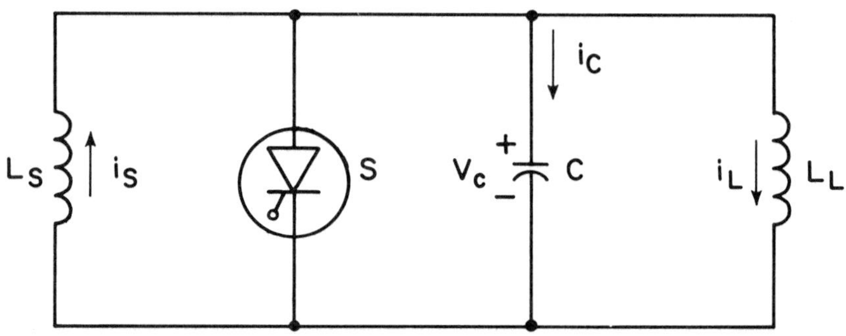

FIG. 2.2. Single-shunt capacitor (LCL) circuit.

2.1.1 Single-Shunt Capacitor (LCL)

A schematic LCL circuit diagram is shown in Figure 2.2. The storage and load magnets are L_S and L_L, respectively. These magnets behave as large inductances with L/R time constants approaching infinity (high-Q magnets). An ideal cycle of operation of the circuit is as follows. The storage magnet, L_S, is precharged to a current I_0 by external means, while the load current, i_L, is zero. Switch S is closed to allow a freewheeling path for the storage coil current. Therefore, the initial capacitor voltage is zero. To start the energy transfer, S is turned off. Current i_S will begin to flow into C, charging it to a positive voltage with the polarity convention shown in the figure. This voltage, in turn, will drive current i_L into the load coil, L_L. Once the switch S is opened, energy transfer occurs at the natural resonant frequency of the circuit. Storage current, i_S, will eventually swing to zero, while the load current, i_L, is at its maximum. This is the midpoint in the cycle where all of the system energy is delivered to the load coil. The load current, i_L, will then begin to charge C in the negative voltage direction. This voltage polarity will drive positive current, i_S, into the storage coil, L_S. Finally, the load current, i_L, will swing to zero and the storage current, i_S, will reach its original maximum value of I_0. Figure 2.3 shows the circuit current and voltage waveforms through one complete cycle of energy transfer to and from the load coil. This oscillation of energy between L_S and L_L will ideally go on indefinitely. However, the cycle can be interrupted when all of the energy is back in L_S and C is charged to a small positive voltage. Closing S at this instant will cause i_S to freewheel through S. The small initial positive voltage on C will help bring S into conduction upon a gate command pulse arrival.

The values of current and voltage in the circuit, with the initial current of I_0 in L_S, are

$$i_s(t) = \frac{L_S I_0}{L_S + L_L} + \frac{L_L I_0}{L_S + L_L} \cos\omega_0 t, \tag{2.1}$$

$$i_L(t) = \frac{L_S I_0}{L_S + L_L} - \frac{L_S I_0}{L_S + L_L} \cos\omega_0 t, \tag{2.2}$$

$$i_c(t) = I_0 \cos\omega_0 t, \tag{2.3}$$

$$v_L(t) = v_S(t) = v_C(t) = \frac{I_0}{\omega_0 C} \sin\omega_0 t, \tag{2.4}$$

where

$$\omega_0^2 \triangleq \frac{L_L + L_S}{L_S L_L C} .$$

An important design consideration is the required amount of installed capacitive energy, U_C. The capacitively stored energy, given by $Cv_C^2/2$, is a maximum in equation (2.4) when $\omega_0 t = \pi/2$. This value is

$$W_{C,\max} = \frac{1}{2} C V_{\max}^2 ,$$

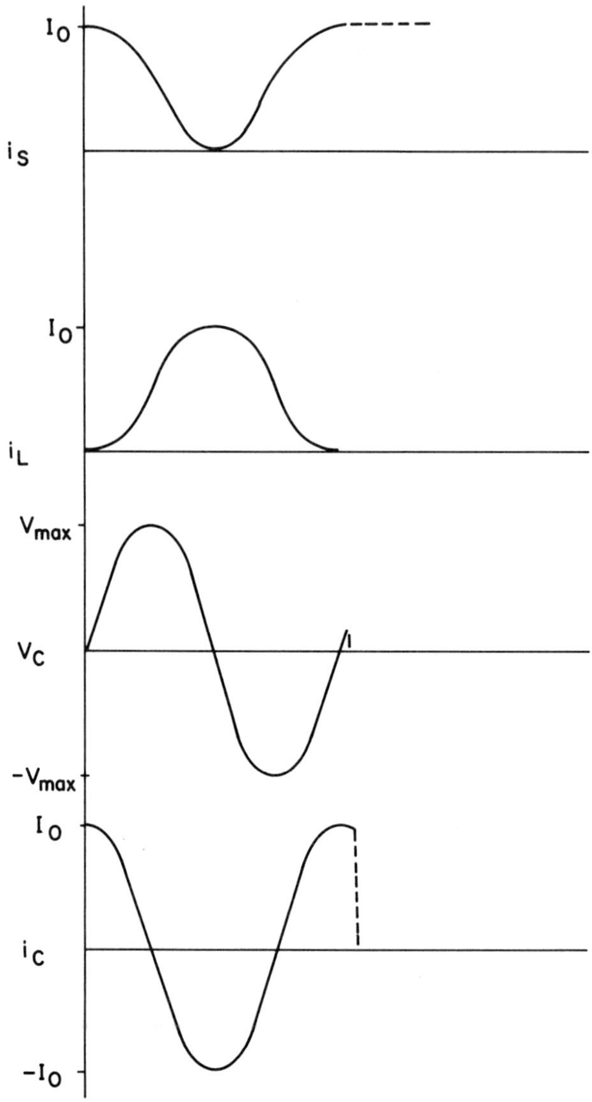

FIG. 2.3. Current and voltage waveforms for an ideal single-shunt capacitor (LCL) circuit in which $L_S = L_L$.

$$W_C = \frac{C}{2} \frac{I_0^2}{\omega_0^2 C^2},$$

$$W_C = \frac{I_0^2}{2 \frac{L_S + L_L}{L_S L_L C} C},$$

$$W_C = \frac{L_L}{L_S + L_L} \left(\frac{1}{2} L_S I_0^2 \right),$$

$$W_C = \frac{L_L}{L_S + L_L} W_0, \tag{2.5}$$

where

$$W_0 \triangleq \frac{1}{2} L_S I_0^2,$$

$$= \text{total stored energy in the system.}$$

When L_L is equal to L_S, the amount of installed capacitive energy is half of the total system energy.

The peak voltage generated and therefore the voltage rating of the switch element is

$$V_{max} = \frac{I_0}{\omega_0 C}. \tag{2.6}$$

The equivalent dc (or time-averaged) voltage, $\langle v \rangle$, to achieve a linear increase of the current in L_L from 0 to I_0 can be found by noting that

$$\langle v \rangle = L \frac{d\langle i \rangle}{dt},$$

where $\langle i \rangle$ is to increase linearly from 0 to I_0 in the energy transfer time corresponding to one-half cycle; i.e., $t = \pi/\omega_0$. Therefore

$$\langle i \rangle(t) = \frac{I_0 \omega_0}{\pi} t$$

and

$$\langle v \rangle = \frac{I_0 \omega_0 L}{\pi}. \tag{2.7}$$

When $L_L = L_S = L$, the resonant frequency, ω_0, is equal to $\sqrt{(2/LC)}$, for which condition the ratio of maximum to average system voltage is

$$\frac{V_{max}}{\langle v \rangle} = \frac{I_0/\omega_0 C}{I_0 \omega_0 L/\pi},$$

$$\frac{V_{max}}{\langle v \rangle} = \frac{\pi}{\omega_0^2 LC},$$

$$\frac{V_{max}}{\langle v \rangle} = \frac{\pi}{2}. \tag{2.8}$$

The switching element, S, can be realized by GTOs. Thus, the operation starts by applying an appropriate negative turn-off pulse to the gate of the device. However, if SCRs are used for realization of S, an external commutation circuit must be provided for the switch turn-off. Figure 2.4 shows an LCL circuit using an SCR with commutation circuit. The commutation capacitor, C_C, is precharged to V_{C0} by an external supply. Commutation of S is initiated by firing SC. The commutation current is established in the loop consisting of C_C, SC, S, and C. This current will rise up to the forward current of S, which is I_0. At this point S is opened and C_C will continue to discharge in the loop consisting of C_C, SC, and L_C. The values of C_C and L_C should be large enough to provide sufficient reverse voltage duration for S to completely recover. Note that although we have introduced another capacitor, inductor, and switching element to the LCL circuit, their values and power ratings are far below those of the main LCL components.

In a practical LCL system, losses will reduce the amount of delivered energy to L_L and recovered energy back to L_S. System losses can be supplied to L_S between operating cycles or in steady state. In either case only a small power supply is needed, when compared to the peak system power.

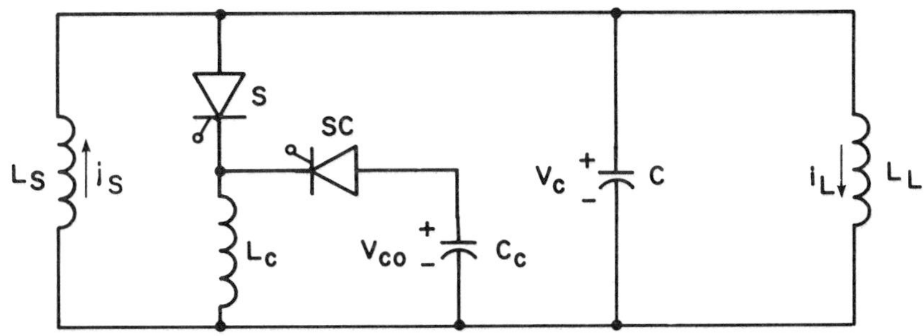

FIG. 2.4. A single-shunt capacitor (LCL) circuit using a commutated SCR switching element.

In the LCL system we have control over the starting time and nonrepetitive operation of the energy transfer. Once the operation begins, no control over the rate of energy transfer can be effected. The rate of energy transfer is specified by the design values of L_S, C, L_L, and I_0. This circuit is inappropriate for large magnetic energy storage applications because up to half of the system energy can appear on the intermediate capacitor. However, it is a candidate for low to medium energy applications where no on-line control of power is required.

2.1.2 Single-Shunt Resistor (LRL)

A schematic LRL circuit is shown in Figure 2.5. The operation of the circuit is as follows. The storage inductor, L_S, is precharged to the initial current, I_0. Switch S is initially closed, and the storage current freewheels through it. Energy transfer is started by turning S off. The storage current will flow through R, dropping a positive voltage on R, as shown in Figure 2.5. This voltage, in turn, will drive a rising current, i_L, in the load coil, L_L. The energy transfer between L_S and L_L in this circuit is neither reversible nor ideally lossless. The exponential nature of energy transfer is described by the following equations of currents and voltages:

$$i_S(t) = \frac{I_0 L_S}{L_S + L_L} + \frac{I_0 L_L}{L_S + L_L} e^{-\alpha t}, \tag{2.9}$$

$$i_R(t) = I_0 e^{-\alpha t}, \tag{2.10}$$

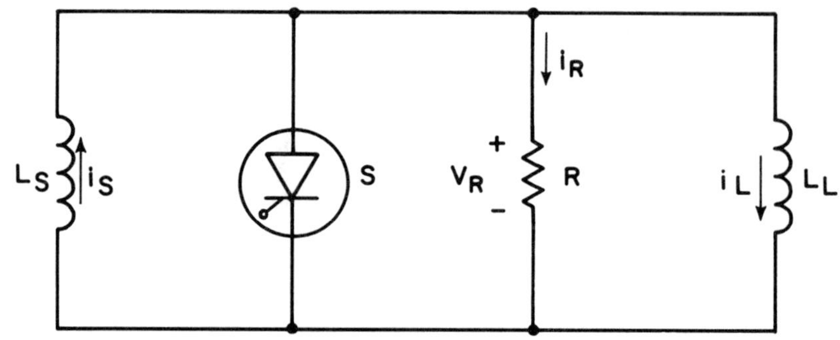

FIG. 2.5. A single-shunt resistor (LRL) circuit.

$$i_L(t) = \frac{I_0 L_S}{L_S + L_L} - \frac{I_0 L_S}{L_S + L_L} e^{-\alpha t}, \tag{2.11}$$

$$v_S(t) = v_L(t) = v_R(t) = I_0 R e^{-\alpha t}, \tag{2.12}$$

where

$$\alpha \triangleq \frac{R}{L_S} + \frac{R}{L_L}.$$

when $t \rightarrow \infty$, the currents in L_S and L_L are equalized to the value $L_S I_0/(L_S + L_L)$. This implies that the residual energy in L_S is

$$W_S(\infty) = \frac{1}{2} L_S \left(\frac{I_0 L_S}{L_S + L_L} \right)^2,$$

$$W_S(\infty) = \frac{L_S^2}{(L_S + L_L)^2} W_0. \tag{2.13}$$

The final energy delivered to L_L is

$$W_L(\infty) = \frac{1}{2} L_L \left(\frac{I_0 L_S}{L_S + L_L} \right)^2,$$

$$W_L(\infty) = \frac{L_S L_L}{(L_S + L_L)^2} W_0. \tag{2.14}$$

The energy dissipated in the resistor is

$$W_R = \int_0^\infty I_0^2 R e^{-2\alpha t} dt$$

$$W_R = \frac{I_0^2 R}{2\alpha},$$

$$W_R = \frac{I_0^2}{2}\left(\frac{L_S L_L}{L_S + L_L}\right),$$

$$W_R = \frac{L_L^2 + L_S L_L}{(L_S + L_L)^2} W_0,$$

$$W_R = \frac{L_L}{(L_S + L)} W_0. \qquad (2.15)$$

It is interesting to note that eventually a stable direct current will circulate in the $L_S L_L$ loop, while the resistor current has decayed to zero. Maximum energy transfer occurs when the two inductances are equal. In this case one-fourth of the system energy is delivered to L_L, one-fourth remains in L_S, and one-half is dissipated in R.

The switch S can be realized by GTOs and SCRs. The commutation circuit for the SCR is similar to that of Figure 2.4. However, higher commutation voltages will be necessary because of the high v_R value as i_S is driven from S to R.

The LRL circuit is useful where only transfer initiation timing control is needed. The rate of energy transfer is preset by the design values of L_S, R, L_L, and I_0. The simplicity of the LRL circuit is exploited in applications where the functions of the switch, S, and resistor, R, are combined in one superconducting switch. The closed superconducting switch is driven to its normal resistive state by a heat pulse. Thus, the commutation or gate turn-off circuitry are replaced by a simpler pulsed heating circuit.

2.2 Two-Switch Circuits

The addition of a second switching element to the energy transfer circuit topology can produce several new circuits. Some of these circuits will have significantly improved controllability. Two dual-switch converter circuits are discussed in this section.

2.2.1 Dual-Switch Single-Shunt Capacitor (S^2LCL) Circuit

A schematic S^2LCL circuit diagram is shown in Figure 2.6. The operation of this circuit is identical to that of the LCL circuit. However, switch S_2 can now be closed when the storage coil is fully discharged ($i_S = 0$) and the load coil is fully charged ($i_L = i_{L,\max}$). Thus, the load coil current will freewheel through S_2 rather than continue to recharge C in the reverse direction, as was described for the LCL circuit. In practice, C should be recharged to a small reverse voltage to facilitate the conduction of S_2 upon arrival of a firing pulse, as shown in Figure 2.7. Energy recovery from L_L to L_S is initiated by turning switch S_2 off. When energy transfer from L_L to L_S is completed, switch S_1 is closed. This completes one full cycle of energy transfer and the circuit is now reset for a new cycle. Figure 2.7 shows the voltage and current waveforms of the S^2LCL circuit.

The switching elements S_1 and S_2 can be realized by GTOs or SCRs. If SCRs are used, individual commutation circuits must be provided for each SCR, as shown in Figure 2.8. The commutation capacitors are precharged by an external power supply prior to each commutation action. The energy losses present in an actual S^2LCL circuit can be replenished in L_S at a low steady-state level or during the intervals between transfer cycles.

The inclusion of the second switch in an S^2LCL circuit provides control over the duration of full energy in the load coil. However, no

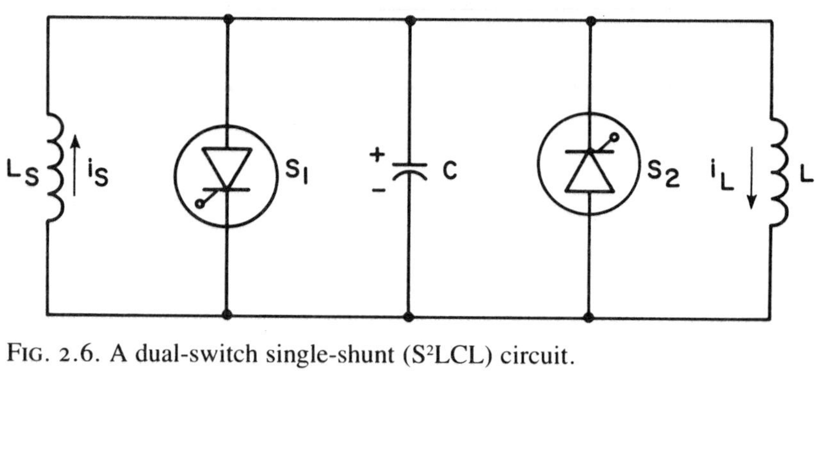

FIG. 2.6. A dual-switch single-shunt (S^2LCL) circuit.

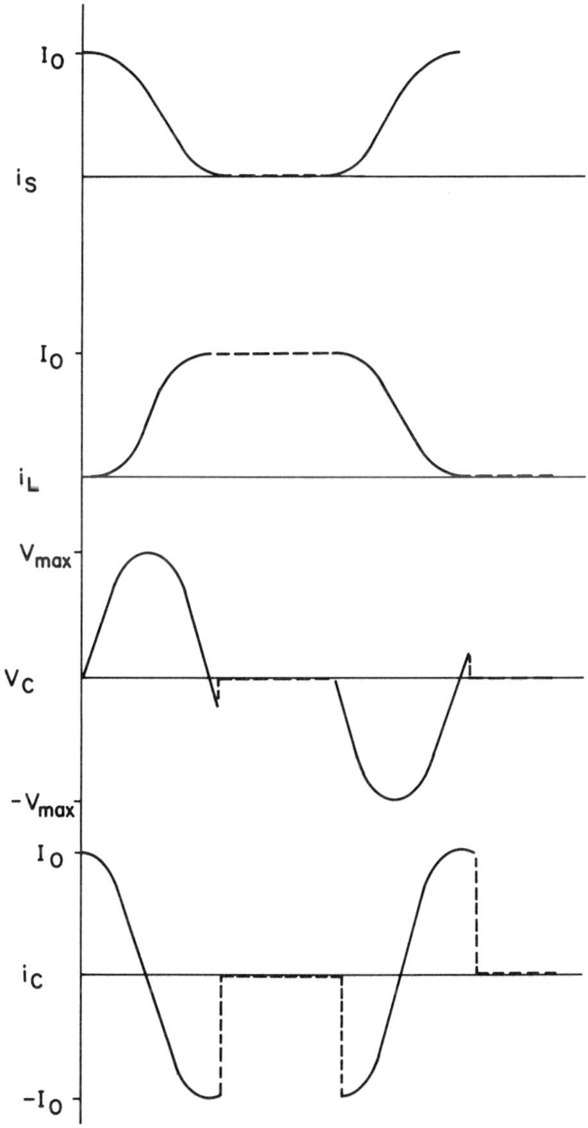

FIG. 2.7. Circuit and voltage waveforms for an ideal dual-switch single-shunt capacitor (S^2LCL) circuit in which $L_S = L_L$.

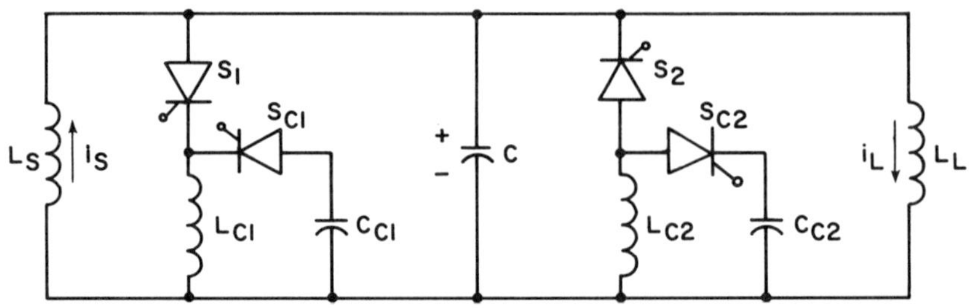

FIG. 2.8. A dual-switch shunt (S^2LCL) circuit using commutated SCR switching elements.

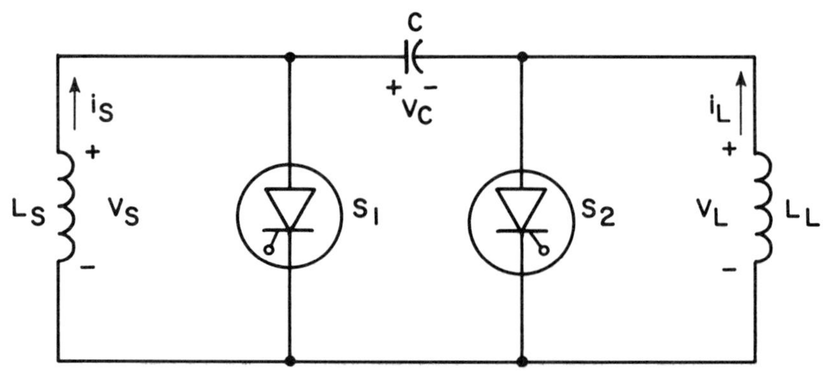

FIG. 2.9. A single flying capacitor (SFC) circuit.

control exists over the rate of energy transfer or intermediate-load magnetic field durations. For such real-time control of the energy transfer, the following two-switch converter may be useful.

2.2.2 Single Flying Capacitor (SFC) Bridge

A schematic SFC circuit diagram is shown in Figure 2.9. This circuit was originally proposed by Simon and Bronner of Princeton Plasma Physics Laboratory. They used ignitrons to realize the switching elements in the experiments. The first experimental SFC circuit using SCRs for the switches was reported by Dick and Dustmann.

In the SFC circuit the total energy stored in the capacitor at any instant is kept to a very small fraction of the total inductively stored energy. This small proportion is achieved by switching current conduction back and forth between S_1 and S_2, which alternately takes and delivers energy between the capacitor and the coils. It follows that the coil currents do not change very much over the switching period, and thus, over a switching time domain, the two coils act as virtual constant current sources. A new energy transfer is possible in this process, and the magnitudes of the two coil currents vary adiabatically with time.

The operation of the circuit during a typical converter cycle is as follows: When S_1 is conducting and S_2 is not, the capacitor is being charged from or discharged into the load coil, depending on the polarity of its initial charge, and the current in L_S is bypassed through S_1. When S_2 is conducting and S_1 is not, the capacitor is being charged from or discharged into the storage coil, depending on the initial charge on the capacitor, and the load coil current is bypassed through S_2.

The voltage waveforms for a typical cycle with $i_S > i_L$ are shown in Figure 2.10. Since the two coils act as current sources over the switching period, the capacitor voltage waveform is a combination of piecewise continuous-line segments. The negative slope corresponds to the interval when S_1 is conducting, and the load coil current discharges the capacitor from $+V_0$ to $-V_{min}$. The positive slope corresponds to the interval when S_2 is conducting and the storage coil current charges the capacitor from $-V_{min}$ to $+V_0$. The charge time is shorter than the discharge time because i_S is greater than i_L.

Note that when S_1 is conducting, v_S is zero, and when S_2 is conducting, v_L is zero. The polarities of the two respective coil voltages cause i_S to slowly decrease, extracting energy from L_S, and i_L to slowly increase, adding energy to L_L.

The right conditions exist for commutation to occur between S_1 and S_2 in the S_1, S_2, and C loop as described in section 1.4.3. The capacitor voltage is at $+V_0$, with the polarity as indicated in Figure 2.9 when the S_2 to S_1 commutation is to take place, and it is at $-V_{min}$ when the S_1 to S_2 commutation is to take place. Figure 2.10 shows the voltage waveforms for several such cycles. The above method of operation will cause maximum rate of energy transfer to the load coil, L_L, if the capacitor, C, is charged to the maximum safe level in every converter cycle. To transfer energy at a rate below the maximum, we can extend the duration of S_1 conduction in the above converter cycle. Thus, the capacitor will be charged to a larger negative voltage from the load coil current, i_L. When S_2 is subsequently closed and S_1 opened, the reverse capacitor voltage must first be discharged by the storage coil current, i_S, before the positive voltage buildup begins. For

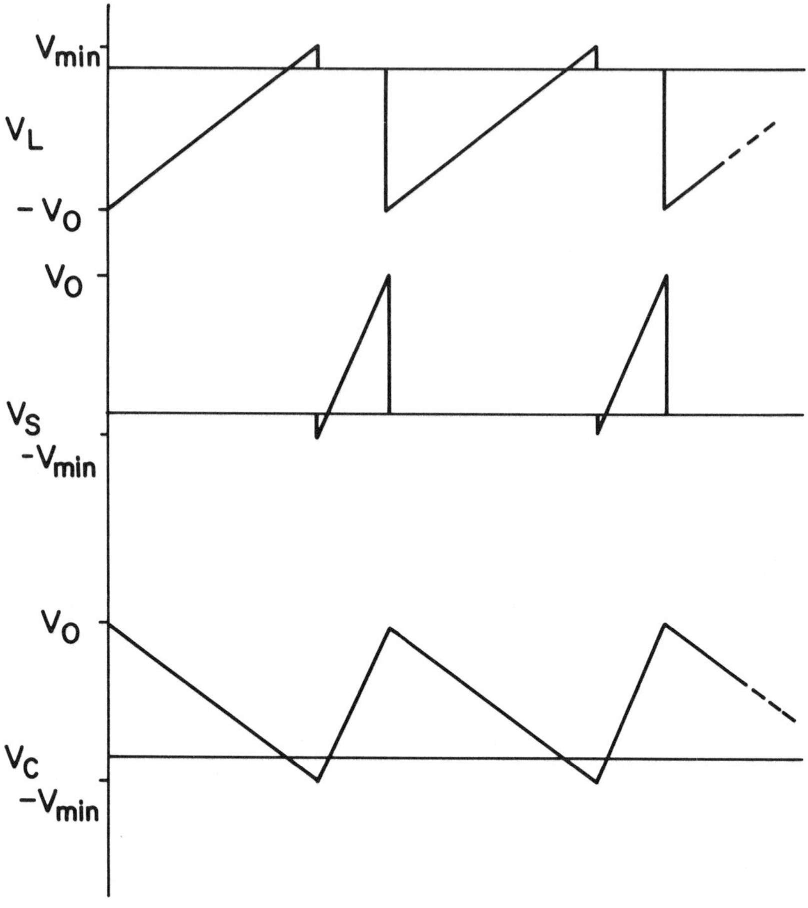

FIG. 2.10. Voltage waveforms for the single flying capacitor (SFC) circuit when $i_S > i_L$.

example, to operate an SFC circuit at zero average power per converter cycle, the conduction times of S_1 and S_2 are set so that the capacitor voltage, v_C, reaches equal positive and negative maximum values. This will cause the capacitor to deliver and absorb equal amounts of energy from each coil per converter cycle; i.e., zero net energy transfer. For other rates of energy transfer in either direction, the positive and negative capacitor voltage peaks are adjusted by controlling the corresponding S_1 and S_2 conduction times. This and other methods of SFC bridge control are discussed in more detail in the next chapter.

The capacitor in an SFC circuit serves two essential functions. It is the intermediate energy buffer between the storage and load coils. In this role it takes energy from one coil and delivers it to the other, with some residual energy usually being returned to the original coil. This capacitor also provides the commutation voltages to the switching elements.

The frequency of SFC converter switching is set sufficiently high so that in every converter cycle only a small fraction of the system's energy is stored in the capacitor. This will substantially reduce the installed capacitive energy requirement. It will also provide fine control over the direction and rate of energy transfer. A complete transfer consists of many thousands of converter cycles. Control can be exercised on each converter cycle throughout the transfer.

The switching elements of an SFC can be simply realized by SCRs. Commutation of the SCRs is readily provided by the SFC circuit topology. However, special SCR turn-off provisions should be provided for the first converter cycle of operation. At the beginning of the operation, i_S is at its maximum value, i_L is equal to zero, the capacitor is at zero voltage, and S_1 is conducting. Firing S_2 will not turn S_1 off because no commutation voltage is present on the capacitor. A similar condition can also occur with all of the system energy trapped in the load and S_2 conducting. Two design approaches can be used for the initial cycle commutation. Each SCR can be provided with its own commutation circuit, as shown in Figure 2.11. Alternatively, capacitor C can be precharged to a minimum required negative voltage by an external means to commutate S_1.

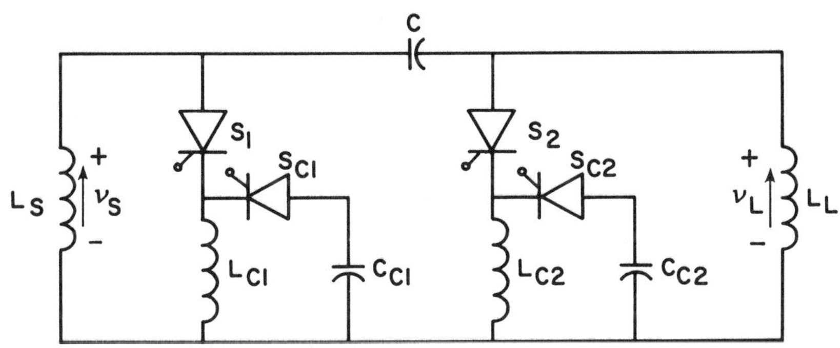

FIG. 2.11. A single flying capacitor (SFC) circuit using commutated SCR switching elements.

The detailed analysis of an SFC circuit is given in the next chapter.

2.3 Four-Switch Circuits

The incorporation of four switching elements in the magnetic energy transfer circuit can produce topologies with new performance and control capabilities. Although several different four-switch converter topologies may be possible, two such circuits are described below.

2.3.1 Four-Switch Single-Shunt Capacitor (S^4LCL) Circuit

A schematic S^4LCL circuit diagram is shown in Figure 2.12. This circuit is briefly described here. The S^4LCL circuit adds the capability of real-time control of power to the original LCL and S^2LCL circuits. The operation of the circuit during a typical converter cycle, while the system energy is divided between L_S and L_L, is as follows: Switches S_1 and S_2 are closed and provide freewheeling paths for the currents i_S and i_L. Switches S_3 and S_4 are open. The capacitor, C, is charged to a negative voltage, based on the polarity convention shown in Figure 2.12. Closing S_3 commutates S_1 off in the commutation loop C–S_3–S_1. S_4 is not conducting, and C is being charged to a positive voltage in the loop C–S_3–L_S by i_S, while i_L is still freewheeling through S_2. When v_C reaches the desired positive voltage, S_1 is closed again, commutating S_3 off in the loop C–S_1–S_3. S_1 and S_2 are now conducting, and both the coil currents are freewheeling while C is holding a positive voltage. Closing S_4 commutates S_2 off in the loop C–S_4–S_2, while i_S is still

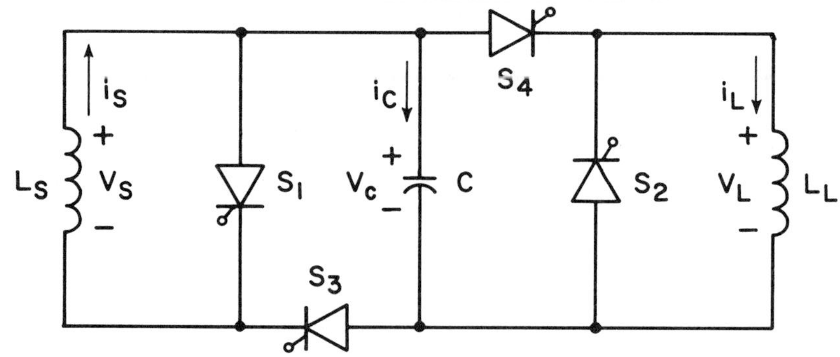

FIG. 2.12. A four-switch single-shunt capacitor (S^4LCL) circuit.

freewheeling through S_1. This switching pattern is maintained until C has completely discharged and is recharged to a desired negative voltage. Closing S_2 commutates S_4 off in the loop C–S_2–S_4. S_2 and S_1 are now conducting, and C is holding a negative voltage. This completes one converter switching cycle and the sequence starts again. The above switching sequence can be summarized as S_3S_2, S_1S_2, S_4S_1, S_2S_1, S_3S_2, . . . where the incoming switch is written first in each conduction pattern.

The frequency of converter switching is set sufficiently high that only a minute fraction of the system energy is stored and transferred by the capacitor in each converter cycle. Thus, the installed capacitive energy storage requirement is kept small. This operation is in contrast to the simple LCL circuit, in which the current follows a complete half-sinusoidal cycle into the capacitor. The short switching periods in S^4LCL circuit periods change the inductor energies and currents by such small amounts that the inductors can be approximated by constant direct current sources in each cycle. Typical voltage and current waveforms of S^4LCL are shown in Figure 2.13. The voltage waveforms are almost linear line segments because of constant inductor current approximation.

The rate of energy transfer is controlled by modulating the S_3S_2 and S_4S_1 conduction pattern durations. This will control the maximum positive and negative capacitor voltages, respectively. In the circuit diagram of Figure 2.12, the energy for positive capacitor voltage is extracted from L_S, and the energy for negative capacitor voltage is extracted from L_L. Therefore, the controller can set the S_3S_2 and S_4S_1 conduction durations for the desired amount and direction of energy extraction per converter cycle. Maximum rate of energy transfer from L_S to L_L is accomplished by charging C to the maximum positive voltage (system voltage rating) and minimum negative voltage (commutation voltage requirement). The converse is true for maximum power from L_L to L_S. The freewheeling intervals S_2S_1 and S_1S_2 are not essential for the operation but may be used for control and operating algorithm optimization. For example, the total converter cycle period, T, can be kept constant for all values of i_S and i_L. The necessary change of S_3S_2 and S_4S_1, as i_S and i_L change through the transfer, are balanced by changes in S_2S_1 and S_1S_2 to maintain a constant period, T.

The zero energy transfer mode in S^4LCL is a quiescent S_2S_1 or S_1S_2 conduction pattern. The same mode in SFC requires full con-

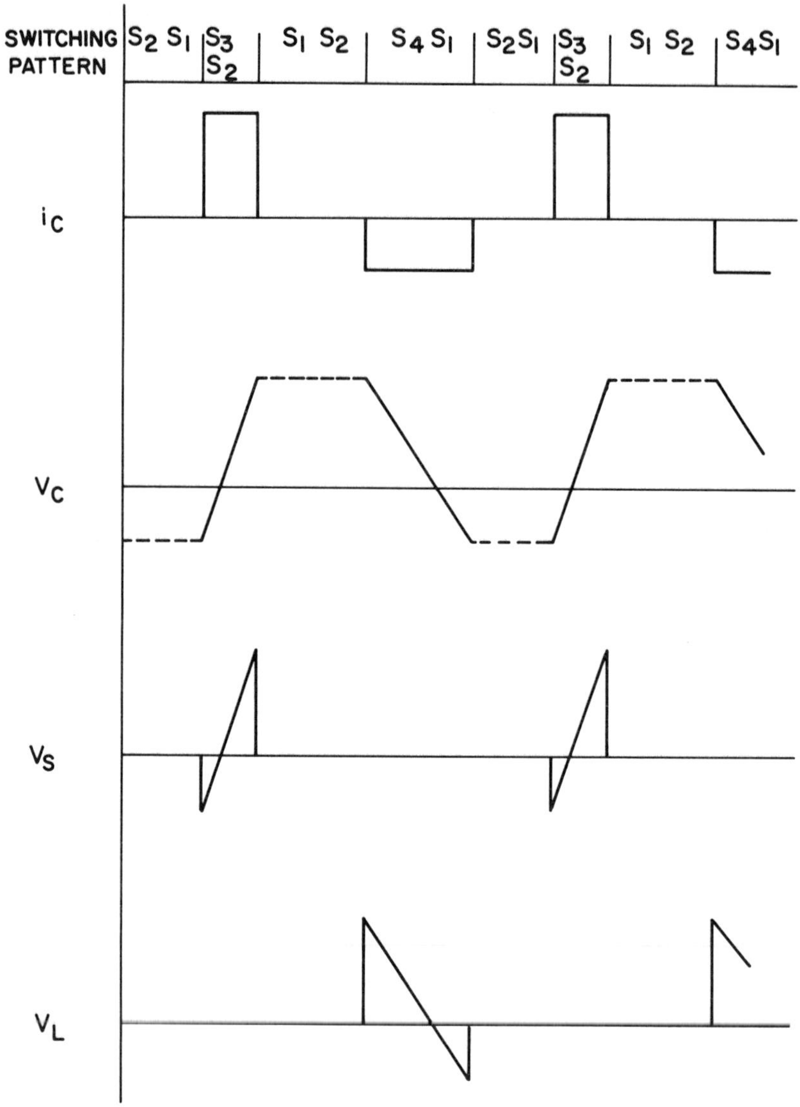

FIG. 2.13. Voltage waveforms for the four-switch single-shunt capacitor (S^4LCL) circuit when $i_S > i_L$ and net energy is being transferred to L_L.

verter switching with its associated switching losses and load voltage ripple. Therefore, for applications where frequent or "quiet" dc-load magnetic fields are required, S^4LCL may be more appropriate and energy efficient. The conduction losses in a practical S^4LCL circuit will cause some decay of i_L in the quiescent mode. However, this change of current may be acceptable for the duration of interest. Furthermore, i_L can be kept to within a specified regulation by occasional S^4LCL converter switching.

As was mentioned in the description of S^4LCL circuit operation, with an initial precharging of the capacitor, C, the circuit will commutate the switching element as required throughout the transfer cycle. Therefore, all the switching elements can be realized by SCRs. However, if an external precharging of C is not allowed, switches S_1 and S_2 should have self-turn-off capability for the first converter switching of the transfer in either direction. In this case S_1 and S_2 can be realized by either GTOs or force-commutated SCRs, while switches S_3 and S_4 are still realized by SCRs.

2.3.2 Dual Flying Capacitor (DFC) Bridge

A schematic DFC circuit diagram is shown in Figure 2.14. The DFC circuit was suggested by Dick and Dustmann. The total energy in the two capacitors of the DFC circuit is kept to a small fraction of the total inductive energy by the fast rate of switching of the SCRs, as is done in the SFC. In fact, as will be shown shortly, the DFC circuit can be viewed as the superposition of two back-to-back SFC circuits. The coils again act as very slowly changing current sources which can be treated as constant over a converter switching period. One of the two upper switches, S_1 or S_2, and one of the two lower switches, S_3 or S_4, are always conducting, so the continuity of the two coil currents is maintained. The evolution from two back-to-back SFC circuits to a single DFC circuit is described in Figure 2.15 and the following discussion.

Consider the two independent SFC circuits shown in Figure 2.15(a). The inductors in these circuits are assumed to be so large that their changes of currents are negligible in one converter cycle. Therefore, these inductors can be visualized as constant current sources for the converter cycle of interest. Connecting these two SFC circuits by the connection x will have no effect on their operation, and no current

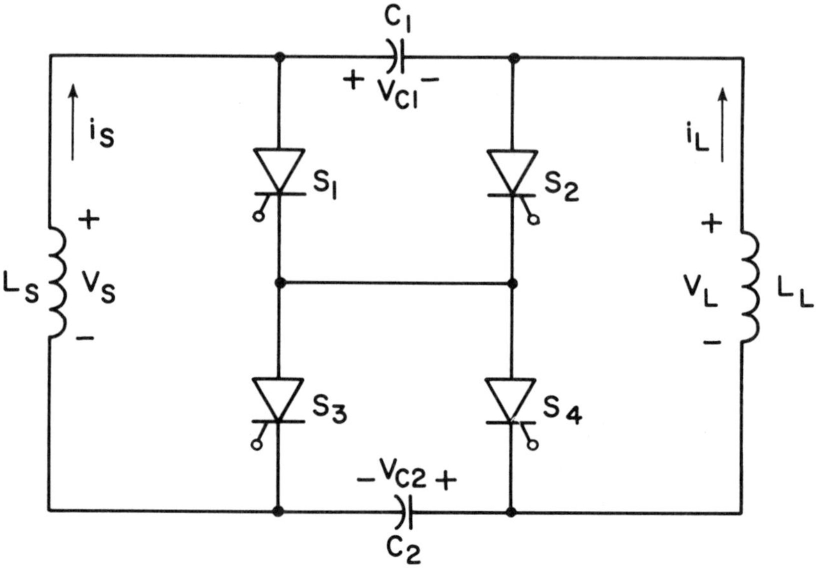

FIG. 2.14. A dual flying capacitor (DFC) circuit.

will flow through x. Note, however, that currents do exist in connections x' and x''. Therefore, the circuit shown in Figure 2.15(b) is completely equivalent to that of Figure 2.15(a). If the two SFC circuits of Figure 2.15(b) are operated with the constraint that the two $L_S/2$ inductors and the two $L_L/2$ inductors have the respective currents i_S and i_L, then the currents in x' and x'' conductors will be identically equal to $i_S + i_L$. Furthermore, no net current will flow through y and z conductors. Therefore, connections y and z can be eliminated without any change in the operation of the two SFC circuits. Equality of currents in the two $L_S/2$ and the two $L_L/2$ inductors can be insured by replacing each pair by their equivalent L_S or L_L inductor. This leads to the circuit of Figure 2.15(c), which is the DFC circuit. Note that the current in conductors x' and x'' is always $i_S + i_L$, regardless of which switch in the upper SFC (S_1 or S_2) and lower SFC (S_3 or S_4) is in the *on* state. The constraint of merging the corresponding inductors of the two SFC circuits limits their independent switching capabilities. Therefore, the DFC circuit is a special case of two independent SFC circuits in which the corresponding storage and load coil currents are allowed to vary in unison.

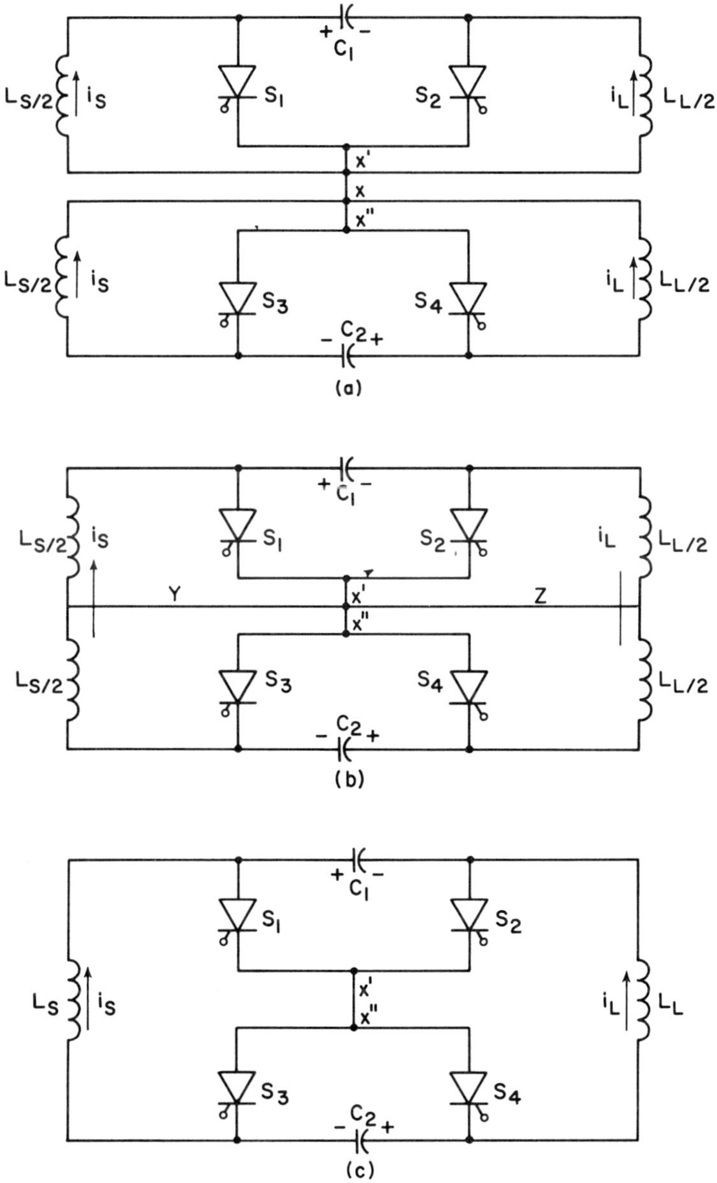

FIG. 2.15. Synthesis of DFC circuit from two SFC circuits.

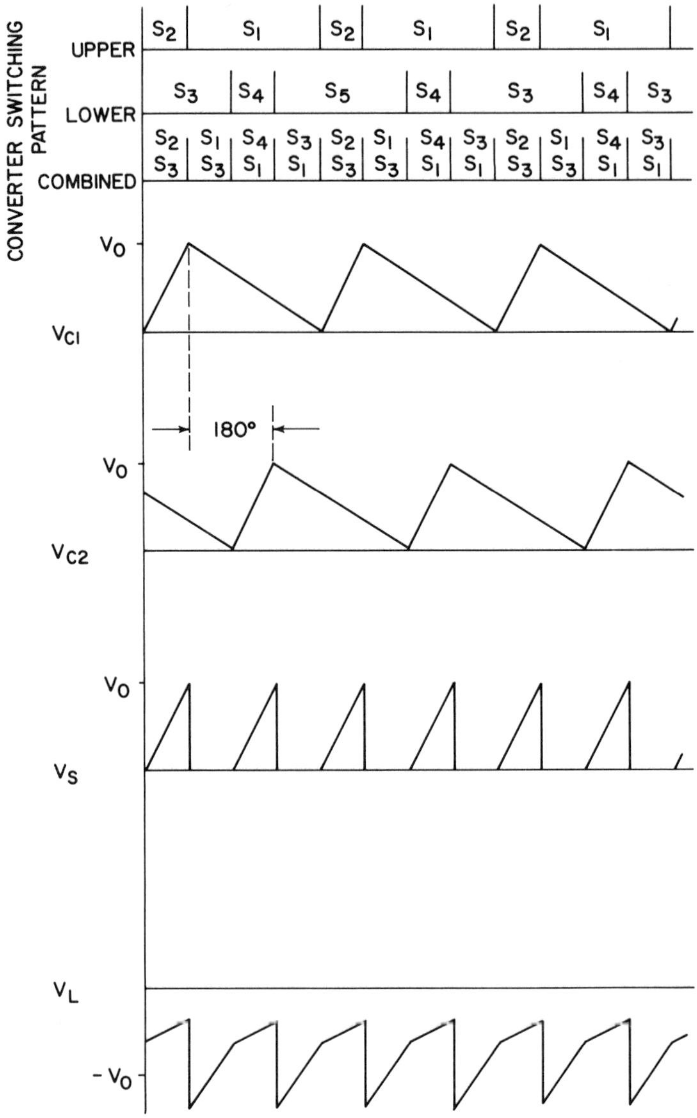

FIG. 2.16. Voltage waveforms for the dual flying capacitor (DFC) circuit when $i_S > i_L$ and net energy is being transferred to L_L.

Note that in the DFC circuit of Figure 2.15(c) commutation between S_1 and S_2 and between S_3 and S_4 is still possible, as is the case for two independent SFC circuits. The polarity on C_1 controls when the S_1–S_2 or S_2–S_1 commutation can or cannot take place. Similarly, the polarity on C_2 controls when the S_3–S_4 or S_4–S_3 commutation can or cannot take place. The voltage waveforms on the two capacitors are identical to an SFC waveform. The voltage waveforms across the two inductors are the sum of the waveforms generated by the upper and lower SFC circuits as shown in Figure 2.10. However, the upper-half and lower-half waveforms will not likely coincide.

Control of power in DFC circuits is accomplished by controlling the individual SFC converters as was described in section 2.2.2. The simplest mode of DFC control is when both SFC converters are operated at the same frequency. This will generate identical voltage waveforms of the two capacitors which are at arbitrary phase difference. An example of waveforms for this control mode at full power from L_S to L_L is shown in Figure 2.16. A more detailed analysis of operation and control of DFC is given in the next chapter.

Commutation of the switching elements of each SFC component of DFC is accomplished by the corresponding capacitor voltage, as was described in section 2.2.2. Here also the commutation of the first converter switching in the energy transfer requires provisions for precharging of the capacitors or forced commutation or gate turn-off. Therefore, either SCRs or GTOs may be used to realize the four switching elements of the DFC.

References

Dick, E. P., and C. H. Dustmann. "Inductive Energy Transfer Using a Flying Capacitor." In *Energy Storage, Compression, and Switching*. Ed. W. H. Bostick, V. Nardi, and O. S. F. Zucker. New York: Plenum Press, 1976.

Gentry, F. E.; F. W. Gutzwiller; N. Holonyak, Jr.; and E. E. Von Zastrow. *Silicon Controlled Rectifiers: Principles and Applications of p-n-p-n Devices*. Englewood Cliffs, N.J.: Prentice Hall, 1964.

Herlet, A., and P. Voss. "State of the Art in Power Semiconductor Devices." Paper presented at IEEE/IAS International Semiconductor Power Converter Conference, Lake Buena Vista, Fla., 1977.

Komarek, P., and A. Ulbright. "Investigations on Superconducting Energy Storage Systems Concerning Fusion Technology." In *Proceedings of 5th International Conference on Magnet Technology*, 1975.

Laquer, H. L. In *Energy Storage, Compression, and Switching*. Ed. W. H. Bostick, V. Nardi, and O. S. F. Zucker. New York: Plenum Press, 1976.

Rioux, C. "Theorie simplifée des procédés de transfert inductif à haut rendement." *Revue de Physique Applique* 10, no. 2 (March 1975).

Rogers, J. D.; C. E. Swannuck; K. I. Thomassen; and D. M. Weldon. *Magnetic Energy Transfer and Storage (METS) Program Schedules for a Fusion Test Reactor*. Report No. LA-5748-Ms. Los Alamos, N.M.: Los Alamos Scientific Laboratory, 1974.

Simon, E. D., and G. Bronner. "An Inductive Energy Storage System Using Ignitron Switching." *IEEE Transactions on Nuclear Science*, n.s., 14, no. 5 (October 1967).

———. *A One Megajoule Inductive Energy Storage System Using Ignitron Switching*. Report No. MATT-741. Princeton, N.J.: Princeton University Plasma Physics Laboratory, 1970.

Thomassen, K. I. "An Inductive Energy Storage System with Capacitive or Homopolar Transfer." In *Proceedings of 5th Symposium on Engineering Problems of Fusion Research*. IEEE Publication No. 73CH0843-3-NPS. Princeton University, 1974.

Ulbricht, A. *Dynamic Behavior of a Resistive Transfer Circuit with a Superconducting Switch*. 9th SOFT, Garmisch-Partenkirchen, Germany, June 14–18, 1976.

CHAPTER 3

Analysis of Nonuniform Switching Circuits: Flying Capacitor Bridges

In this chapter more detailed analytical expressions for the flying capacitor circuits are derived. Furthermore, real-time control of power in these circuits is examined more closely. The expressions and concepts developed will not only add insight but also help in the design of these circuits and their control.

3.1 Analysis of the Single Flying Capacitor (SFC) Bridge

The SFC circuit was introduced in section 2.2.2. The derivation of the circuit voltage and current equations begin when switch S_2 is closed and S_1 is open, as in Figure 2.11.

We will assume that the stored energy in the capacitor at any given time is a small fraction of the total energy in the circuit. With the initial voltage on the capacitor assumed to be zero, the current flow, $i_C(t)$, is

$$i_C(t) = I_0 \cos\omega_S t,$$

where

$$\omega_S = \frac{1}{\sqrt{L_S C}}. \tag{3.1}$$

The capacitor voltage is

$$v_C(t) = \frac{I_0}{\omega_S C} \sin\omega_S t. \tag{3.2}$$

The stored energy in the capacitor, U_C, is

$$U_C = \frac{1}{2} C[v_C(t)]^2;$$

$$U_C = \frac{\frac{1}{2}I_0^2}{C\omega_S^2}\sin^2(\omega_S t);$$

$$U_C = \frac{1}{2}L_S I_0^2 \sin^2(\omega_S t). \tag{3.3}$$

Since the initial energy stored in the circuit is $1/2L_S I_0^2$, $\omega_S t$ must be kept small if U_C is to be a small fraction of the total stored energy in the circuit.

If the initial charging time is denoted as Δt such that $\omega_S \Delta t << \pi/2$, then $v_C(t)$ is

$$v_C(t) \simeq \frac{I_0}{\omega_S C}(\omega_S \Delta t),$$

$$v_C(t) \simeq \frac{I_0 \Delta t}{C}. \tag{3.4}$$

When the voltage on the capacitor reaches the desired level, a gate pulse is delivered to S_1. Voltage and current conditions on the bridge are shown in Figure 3.1. The voltage on C in the closed loop consisting of C, S_1, L_{C1}, S_2, and L_{C2} has polarity to assist in the forward bias of

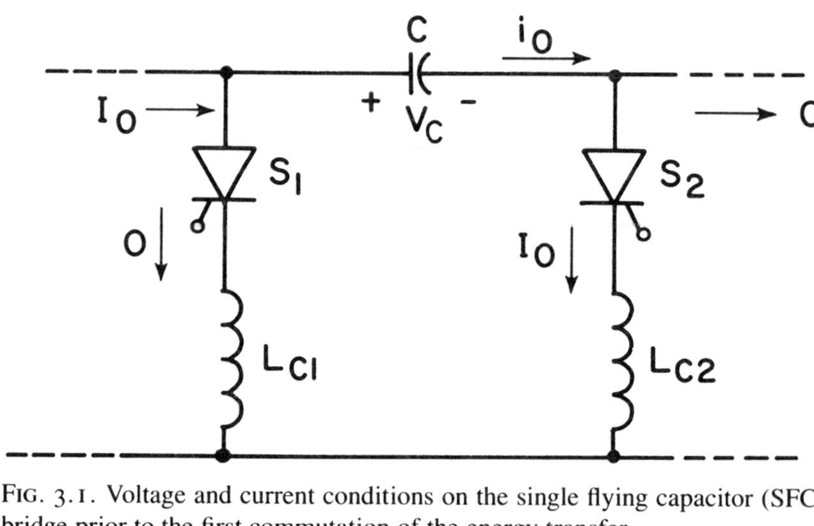

FIG. 3.1. Voltage and current conditions on the single flying capacitor (SFC) bridge prior to the first commutation of the energy transfer.

S_1 and the reverse bias of S_2; i.e., it acts as a counter pulse circuit for the commutation of current from S_2 to S_1.

The commutation duration, t_c, is so short that U_C in equation (3.3) does not vary appreciably. Therefore, $i_S(t)$ remains almost unchanged:

$$i_S(t) = I_0 \cos\omega_S t_C,$$

$$i_S(t) \simeq I_0(1 - \omega_S^2 t_C^2/2);$$

$$i_S(t) \simeq I_0. \tag{3.5}$$

Once the current has switched from S_2 to S_1, the current in L_S is shunted by the thyristor S_1, and voltage on capacitor C is placed across L_L. The voltage and current in the L_L, C, and S_1 loop using equation (3.4) are

$$v_L(t) = -\frac{I_0 \Delta t}{C} \cos\omega_L t, \tag{3.6}$$

and

$$i_L(t) = -\frac{1}{L_L} \int v_L(t) dt; \tag{3.7}$$

where

$$\omega_L = \frac{1}{\sqrt{L_L C}}.$$

Equation (3.7) is transformed by integration into

$$i_L(t) = I_0 \omega_L \Delta t \sin\omega_L t, \tag{3.8}$$

with the boundary condition for $i_L(0)$ equal to 0.

The oscillation described in equation (3.6) is allowed to go slightly beyond a quarter cycle. The energy from C is transferred to L_L, except for a slight amount of energy put into C with a reverse voltage polarity. At this point a gate pulse on S_2 will cause commutation of the current from S_1 to S_2.

Both inductor currents will flow through S_2. The current $i_s(t)$ starts to discharge whatever amount of charge was left in C and then continues to charge C again with positive polarity.

After several energy transfer steps following the sequence described above, the energy in L_L becomes much larger than the stored energy in the capacitor. We will study the circuit behavior under the conditions where $1/2(L_S i_S^2)(t)$ and $1/2(L_L i_L^2)(t) >> 1/2(C v_C^2)(t)$.

We will pick the instant in time, t_i, where

$$v_C(t_i) = V_i, \tag{3.9}$$

$$i_L(t_i) = I_L(t_i) \tag{3.10}$$

and

$$i_S(t_i) = I_S(t_i), \tag{3.11}$$

and a gate pulse is transmitted to S_1. The two coil currents flowing through S_2 are transferred to S_1. The voltage across L_S is clamped to zero by the conduction of S_1. The current $I_S(t_i)$ remains constant during the period S_1 conducts. The voltage on C is placed across L_L, opposite to the polarity mark indicated in Figure 2.11. The current and voltages following t_i by the amount of time Δt later are

$$v_S(t_i + \Delta t) = 0, \tag{3.12}$$

$$i_S(t_i + \Delta t) \simeq I_S(t_i), \tag{3.13}$$

$$v_C(t_i + \Delta t) = V_i \cos(\omega_L \Delta t) - I_L(t_i)\omega_L L_L \sin(\omega_L \Delta t), \tag{3.14}$$

$$v_L(t_i + \Delta t) = -v_C(t_i + \Delta t), \tag{3.15}$$

and

$$i_L(t_i + \Delta t) = I_L(t_i)\cos(\omega_L \Delta t) + \frac{V_i}{\omega_L L_L}\sin(\omega_L \Delta t). \tag{3.16}$$

Defining the time Δt, for when the capacitor voltage goes to zero, as t_d,

$$\tan(\omega_L t_d) = \frac{V_i}{\omega_L L_L I_L(t_i)};$$

$$\tan(\omega_L t_d) = \frac{\sqrt{C}V_i}{\sqrt{L_L}I_L(t_i)}. \tag{3.17}$$

The right side of equation (3.17) is the square root of the ratio of capacitive energy to inductive energy in L_L. This is a small number by definition, and so the tangent is a small number; i.e.,

$$\tan(\omega_L t_d) \simeq \omega_L t_d \tag{3.18}$$

for any $\Delta t \leqslant t_d$, $\omega_L \Delta t$ is also small, and we can approximate

$$\sin(\omega_L \Delta t) \simeq \omega_L \Delta t, \tag{3.19}$$

and

$$\cos(\omega_L \Delta t) \simeq 1 - \frac{\omega_L^2 (\Delta t)^2}{2}. \tag{3.20}$$

Equations (3.12) and (3.16) reduce to

$$v_S(t_i + \Delta t) = 0, \tag{3.21}$$

$$i_S(t_i + \Delta t) = I_S(t_i), \tag{3.22}$$

$$v_C(t_i + \Delta t) = V_i - \frac{I_L(t_i)\Delta t}{C}, \tag{3.23}$$

$$v_L(t_i + \Delta t) = -V_i + \frac{I_L(t_i)\Delta t}{C}, \tag{3.24}$$

$$i_L(t_i + \Delta t) = I_L(t_i) - I_L(t_i)\left[\frac{\omega_L^2 (\Delta t)^2}{2}\right] + \frac{V_i \Delta t}{L_L}. \tag{3.25}$$

Equation (3.25) can be further reduced since

$$I_L(t_i)\left[\frac{\omega_L^2 (\Delta t)^2}{2}\right] = I_L(t_i)\frac{(\Delta t)^2}{L_L C 2};$$

$$I_L(t_i)\left[\frac{\omega_L^2 (\Delta t)^2}{2}\right] = \frac{V_i \Delta t}{2 L_L}. \tag{3.26}$$

The results of equation (3.26) can be used in equation (3.25) to yield

$$i_L(t_i + \Delta t) = I_L(t_i) + \frac{V_i \Delta t}{2 L_L} \tag{3.27}$$

The two inductors, L_L and L_S, behave like two current sources that charge and discharge the capacitor C at the rate of $(I\Delta t/C)$. Currents I_S and I_L change by very small amounts over the switching cycle, not enough to alter the linear behavior of the equations (3.21) to (3.25). A typical graph of the circuit voltages for $I_S = 2I_L$ over several switching cycles is shown in Figure 3.2.

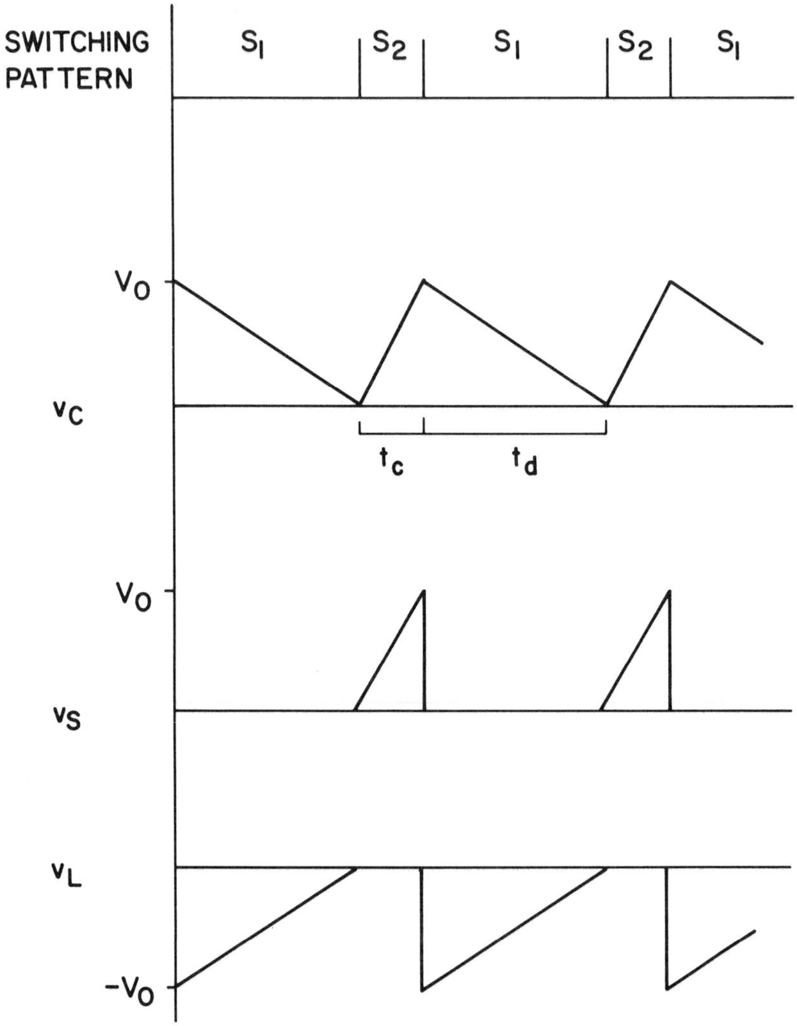

FIG. 3.2. Typical circuit voltages for the single flying capacitor (SFC) when $i_S > i_L$.

The commutation times in an SFC circuit are determined by voltage comparators placed across the capacitor. One voltage comparator is adjusted to send a gate pulse to S_1 when the logic system indicates that V_0 is reached on the capacitor, and the other voltage comparator is adjusted to send a gate pulse to S_2 when the capacitor voltage goes slightly negative.

3.1.1 Voltages, Currents, and Power Flow

The discharge and charge times for the capacitor are denoted in Figure 3.2 by t_d and t_c, respectively. Since the coil currents stay relatively constant over the switching cycle, the times t_d and t_c are given by

$$t_d = \frac{V_0 C}{I_L} \tag{3.28}$$

and

$$t_c = \frac{V_0 C}{I_S}. \tag{3.29}$$

The times t_c and t_d vary continually throughout the energy transfer cycle. The SFC bridge is controlled by voltage comparator circuits, not by clock pulses. The values of t_c and t_d are used as computational aids in the further development of the SFC concept.

The average voltage on the load coil over a switching cycle is:

$$\langle v_L \rangle = \frac{1}{t_c + t_d} \int_0^{t_d} \left(\frac{V_0}{t_d} t - V_0 \right) dt;$$

$$\langle v_L \rangle = \frac{1}{\left(1 + \frac{I_L}{I_S}\right) t_d} \left(\frac{V_0 t^2}{2 t_d} - V_0 t \right) \Bigg|_0^{t_d};$$

$$\langle v_L \rangle = \frac{1}{\left(1 + \frac{I_L}{I_S}\right)} \left(- \frac{V_0}{2} \right);$$

$$\langle v_L \rangle = - \frac{V_0}{2} \left(\frac{I_S}{I_L + I_S} \right). \tag{3.30}$$

Likewise, the average storage coil voltage is:

$$\langle v_S \rangle = \frac{V_0}{2}\left(\frac{I_L}{I_L + I_S}\right). \tag{3.31}$$

At the start of the energy transfer cycle, the average load coil voltage is $-V_0/2$, and the average storage coil voltage is $\simeq 0$. The average power during the early part of the transfer cycle is nearly zero; i.e., the load coil voltage is $-V_0/2$, the load coil current is nearly zero, and the storage coil voltage is nearly zero, even though the coil current is large (see Figure 2.10).

When the currents in the two coils are equal, the average voltages on the two coils are both equal to $V_0/4$. If the current in the storage coil approaches zero, the average voltage on the load coil approaches zero, and the average voltage on the storage coil approaches $V_0/2$.

The electrical losses in the energy transfer system are due to I^2R losses in the cables and potential drop across the switches. If one assumes that these losses can be made negligible, then the energy transfer system is conservative and the following is true:

$$\frac{1}{2} L_S I_S^2(t) + \frac{1}{2} L_L I_L^2(t) = U_0, \tag{3.32}$$

where U_0 is the total energy in the two coils at $t = 0$.

The following parametric forms can be used for the currents $i_S(t)$ and $i_L(t)$:

$$I_S(t) = I_x \cos[\theta(t) + \sigma] \tag{3.33}$$

and

$$I_L(t) = I_x \sqrt{\frac{L_S}{L_L}} \sin[\theta(t) + \sigma]. \tag{3.34}$$

When the currents given in equations (3.33) and (3.34) are substituted into equation (3.32), one gets $1/2(L_S I_x^2) = U_0$ or

$$I_x = \sqrt{\frac{2U_0}{L_S}}. \tag{3.35}$$

Assuming that $\theta(0) = 0$, the value of σ will be

$$\sigma = \arctan\sqrt{\frac{L_L}{L_S}}\left[\frac{I_L(0)}{I_S(0)}\right], \tag{3.36}$$

in order to match the initial conditions.

The currents in the two coils change slowly with time and to first order are given by

$$I_L(t) = -\frac{1}{L_L}\int_0^t \langle v_L(t)\rangle dt, \tag{3.37}$$

and

$$I_S(t) = \frac{1}{L_S}\int_0^t \langle v_S(t)\rangle dt. \tag{3.38}$$

Substitution from equations (3.30), (3.31), (3.33), (3.34), and (3.36) into equation (3.37) and differentiation of equation (3.37) with time leads to an equation for θ.

$$L_L\frac{d}{dt}\left\{I_x\sqrt{\frac{L_S}{L_L}}\sin[\theta(t) + \sigma]\right\}$$

$$= \frac{V_0}{2}\left\{\frac{\cos(\theta(t) + \sigma)}{\sqrt{L_S/L_L}\sin[\theta(t) + \sigma] + \cos[\theta(t) + \sigma]}\right\};$$

$$I_x\sqrt{L_S L_L}\frac{d\theta}{dt}\cos(\theta + \sigma)$$

$$= \frac{V_0}{2}\frac{\cos(\theta + \sigma)}{\sqrt{L_S/L_L}\sin(\theta + \sigma) + \cos(\theta + \sigma)};$$

$$\left\{\sin(\theta + \sigma) + \sqrt{\frac{L_L}{L_S}}\cos(\theta + \sigma)\right\}d\theta = \frac{V_0}{2I_x L_S}dt;$$

$$\sqrt{1 + \frac{L_L}{L_S}} \left\{ \frac{1}{\sqrt{1 + L_L/L_S}} \sin(\theta + \sigma) \right.$$

$$\left. + \frac{\sqrt{L_L/L_S}}{\sqrt{1 + L_L/L_S}} \cos(\theta + \sigma) \right\} d\theta = \frac{V_0}{2I_x L_S} dt;$$

$$\sin(\theta + \sigma + \beta) d\theta = \frac{V_0 dt}{2I_x L_S \sqrt{1 + L_L/L_S}};$$

where

$$\beta = \arctan \sqrt{L_L/L_S}. \tag{3.39}$$

Integration of equation (3.39) leads to

$$\cos\,[\theta(t) + \sigma + \beta] = - \frac{V_0 t}{2I_x L_S \sqrt{1 + L_L/L_S}} + \text{constant}. \tag{3.40}$$

The choice of the constant, σ, in equation (3.36) is shown under the assumption that $\theta(t)$ is equal to zero when t is equal to zero. Therefore, the constant of integration in equation (3.40) must be equal to $\cos(\sigma + \beta)$ to make $\theta(0)$ equal to zero at t equal to zero. The equation for $\theta(t)$ is

$$\cos[\theta(t) + \sigma + \beta] = - \frac{V_0 t}{2I_x L_S \sqrt{1 + L_L/L_S}} + \cos(\sigma + \beta). \tag{3.41}$$

The average power flow between the coils is given by the product of $i_L(t)$ and $v_L(t)$,

$$\langle p \rangle = \frac{V_0 I_x}{2} \frac{\sqrt{L_S/L_L} \cos(\theta + \sigma) \sin(\theta + \sigma)}{\sqrt{L_S/L_L} \sin(\theta + \sigma) + \cos(\theta + \sigma)} \tag{3.42}$$

If the two coils have equal inductance, and all the current at $t = 0$ is in one of the two coils, then it is easy enough to show that the maximum average power occurs when θ is equal to 45°, and the value of maximum power is equal to $V_0 I_0 \sqrt{(2)}/8$ where I_0 is the initial current in the inductor.

3.1.2 Active Control of the Circuit by Coil Power Reversal

Active control of the energy transfer is a design goal of the circuit and in an SFC bridge is simple enough to accomplish. Switching from S_1 to S_2 in the SFC bridge was described in section 2.2.2 as occurring shortly after v_C became slightly negative. In fact, the voltage comparator which detects when v_C goes slightly negative could be controlled to allow the gate pulse directed to S_2 to be sent when capacitor voltage reaches an arbitrary value of voltage from almost zero to $-V_0$. The waveforms that result from changing the gate timing level on the negative side comparator are shown in Figure 3.3. The period for the waveform of Figure 3.3 is

$$T = t_c + t_d + kt_d + kt_c, \tag{3.43}$$

where t_c and t_d are given by equations (3.28) and (3.29) and k is a parameter between 0 and 1. The average value for load coil voltage is

$$\langle v_L(t) \rangle = -\frac{V_0}{2}\left(\frac{I_S}{I_S + I_L}\right)(1 - k). \tag{3.44}$$

Thus, by allowing k to vary from 0 to 1, the time-averaged value of $v_L(t)$ varies from maximum to zero. If one considers the switching pattern for Figure 3.3, where S_2 is in the conducting state for v_C from $-kv_0$ to $+v_0$, and S_1 is in the conducting state from v_0 to $-kv_0$, the average energy flow could be fully reversed by switching from S_2 to S_1 just after v_C becomes slightly positive.

Active control of the SFC bridge can be accomplished with the use of the voltage comparators, linear amplifiers, and zero crossing networks. No sophisticated microprocessors are required. The reference voltage is set by a transfer function that measures the state of system parameters related to the utilization of the load coil. A single bistable amplifier or flip-flop gate controls whether the reference operates on the negative polarity of capacitor voltage or positive polarity as required by the demand for the energy flow into or out of the load coil.

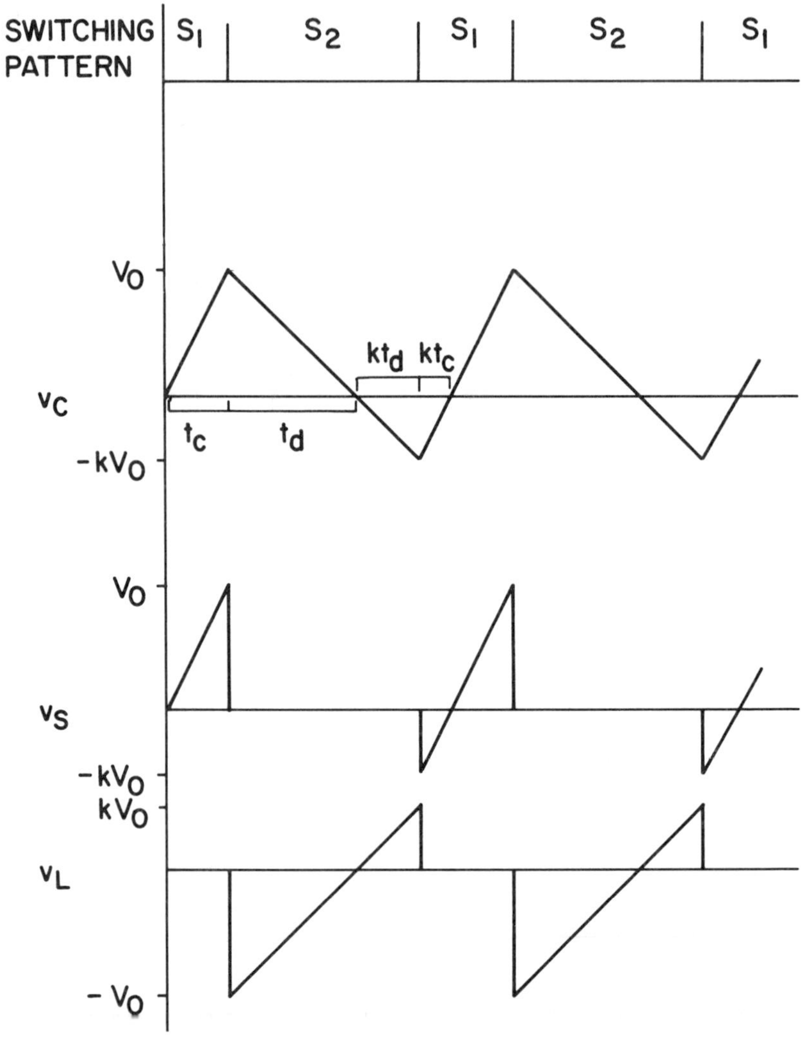

FIG. 3.3. Voltage waveforms in a single flying capacitor (SFC) bridge with active control.

3.1.3 Component Ratings

One design goal is to minimize capacitive energy storage. The voltage limit in the circuit can be imposed by a variety of constraints such as the insulation limit of the inductors or space limitations of the equipment. The economics of capacitive energy storage is also a consideration for setting V_0. Purchase cost per unit energy tends to be lowest in the range of 5 to 15 kV, and although it is possible to stack capacitors in series, installation costs start increasing rapidly because of voltage insulation requirements of the support frame and because of voltage distribution hardware. Given the various constraints, then, the value of V_0 is fixed. The capacitive energy storage, U_C, is

$$U_C = \frac{1}{2}CV_0^2;$$

$$U_C = \frac{1}{2}C\left(\frac{I_0^2 t_c^2}{C^2}\right);$$

$$U_C = \frac{I_0^2 t_c^2}{2C} \tag{3.45}$$

The value of V_0 is $I_0 t_c / C$. Therefore, the value of U_C is directly proportional to the product of I_0, V_0, and t_c. The first two parameters, I_0 and V_0, are defined by the system, and the last parameter, t_c, is defined by the switching device. Thus, to minimize capacitive energy, the switches should operate at the shortest time permitted by the devices used in the switch arms. The turn-off time for current state-of-the-art SCR devices varies between 10 and 100 μs; this means for efficient thyristor operation, a minimum conduction period for t_c or t_d is about 0.5 ms.

The ratio of capacitively stored energy from equation (3.45) to the total stored energy in the circuit, i.e., $1/2(L_S I_0^2)$, has the value $\omega_S^2 t_c^2$. This was by definition a small number, and thus the validity of the assumptions used to go from equation (3.3) to (3.4) can be tested. The other component ratings can now be addressed. The thyristor switches S_1 and S_2 are connected directly across L_L and L_S and therefore are never subjected to any voltages exceeding V_0. Of course, the average control voltage on the load coil is never greater than $V_0/2$ and becomes smaller by the ratio $I_S/(I_L + I_S)$, which in the extreme goes to zero as I_S goes to 0. In fact, $V_0/\langle v_L \rangle$ is

$$\frac{V_0}{\langle v_L \rangle} = -2\left(\frac{I_L + I_S}{I_S}\right). \tag{3.46}$$

Thus, in order to achieve a measure of control as defined by $\langle v_L \rangle$ in the circuit with some given value of I_L, it is necessary to maintain a sufficiently large value of I_S. The constraints on the application of the SFC circuit ultimately define these quantities and in return help determine how high the value V_0 needs to be and what size of L_S is required. It is noteworthy, however, that unless $L_S >> L_L$, the voltage ratings on the thyristors are going to be a sizeable factor larger than the control voltage.

The instantaneous current in the switches S_1 and S_2 are the sums of the inductor currents, $i_S(t)$ and $i_L(t)$. During the energy transfer cycle, the ratio between $i_S(t)$ and $i_L(t)$ is slowly changing. Assuming at some instant in time that $i_S(t) = I_S$ and $i_L(t) = aI_S$, and L_S is equal to L_L, where a is a numerical parameter, the value of I_S is equal to $I_0/\sqrt{[(1 + a)^2]}$. The rms value of the current in the switch element is

$$I_{rms} = \sqrt{\frac{1}{T}\int_0^{t_d}(I_S + aI_S)^2 dt};$$

$$I_{rms} = I_S\sqrt{\frac{t_d}{T}(1 + a)^2} \tag{3.47}$$

Substituting V_0C/aI_S for t_d and $V_0C/aI_S + V_0C/I_S$ for T in equation (3.47) results in the following expression for I_{rms},

$$I_{rms} = I_S\sqrt{1 + a};$$

$$I_{rms} = I_0\frac{\sqrt{1 + a}}{\sqrt{1 + a^2}} \tag{3.48}$$

The maxima for I_{rms} with respect to a can be found from equation (3.48). The value of a that corresponds to the maximum value of I_{rms} and, therefore, the design value of the circuit, is equal to $\sqrt{2} - 1$. The value of equation (3.48) with $a = \sqrt{2} - 1$ is

$$I_{rms} = \frac{I_0}{\sqrt{2\sqrt{2} - 2}};$$

$$I_{rms} \simeq 1.0987 I_0. \tag{3.49}$$

3.1.4 Efficiency of Energy Transfer

Of course, efficiency of energy transfer is an ambiguous entity in a system for which arbitrary and active changes in the transfer cycle are expected. However, some broad comments can be made which will be quantified later when the different circuits are compared.

Losses in the circuit are caused by the room-temperature power leads, the capacitor, and the SCR switches. Joule heating in the power leads is equal to the product of the square of the current density, the volume of the conductor, and the resistivity of the conducting material. Since the length of the conductor is ultimately fixed by the installation design, the losses in the power leads are inversely proportional to the cross-sectional area of the conductor. Power losses in the capacitor are a result of joule heating in the internal leads and dielectric losses. These losses can be made small by increasing the size of the internal leads and decreasing the stress on the dielectric by increasing its volume. All these losses can be made arbitrarily small, but this is done at the expense of initial capital investment, and economics will dictate how far the losses will be decreased in these elements.

The switch losses are a combination of the forward conducting loss and the transient switching losses. Note in Figures 1.4 and 1.5 that when the SCR is in the fully conducting state, a small voltage, V_B, exists. This voltage causes a forward conducting loss that is proportional to the product of it and the current through the SCR. There are also losses in the SCR during the transient periods between fully conducting and fully isolating. These losses can be determined by integrating the product of voltage and current over transient intervals of Figures 1.4 and 1.5. The relevant importance of the transient switching loss depends on the choice of operating period in the forward conducting state. These losses are device dependent, but, typically, if the switching losses are about equal to the forward conduction losses, then the forward conducting period is ten times greater than the switching times of the device. As the conducting period is increased, the losses in the switching interval become relatively less important. Finally, there are losses in the snubber circuits whenever the SCRs must be connected in series (see Figure 1.7).

Ignoring economics, the ultimate energy loss is a result of the forward conduction loss of the SCRs. The forward voltage drop is an inherent feature of the device which cannot be reduced. It is instructive, therefore, to calculate this loss, since it determines the point at which reducing other losses becomes decreasingly effective. Given a system voltage, V_0, then S_1 and S_2 must be rated to withstand that level. If the voltage rating of the thyristor is V_T and the system voltage is V_0, then V_0/V_T number of thyristors, or more, must be put in series. The energy lost in the transfer because of thyristor losses is equal to the time integral of the product of the number of thyristors in series, the barrier voltage drop, and the sum of the coil currents. The current is transferred back and forth between S_1 and S_2, so the total losses are the sum of the losses in S_1 and S_2; i.e.,

$$U_e = \int_0^t \frac{V_0 V_B}{V_T} I_x \left(\cos[\theta(t) + \sigma] + \sqrt{\frac{L_S}{L_L}} \sin[\theta(t) + \sigma] \right) \times \frac{t_d}{t_c + t_d} dt + \int_0^t \frac{V_0 V_B}{V_T} I_x \times \left(\cos[\theta(t) + \sigma] + \sqrt{\frac{L_S}{L_L}} \sin[\theta(t) + \sigma] \right) \times \frac{t_c}{t_c + t_d} dt;$$

$$U_e = \int_0^t \frac{V_0 V_B}{V_T} I_x \sqrt{\frac{L_S}{L_L}} \times \left(\sin[\theta(t) + \sigma] + \sqrt{\frac{L_L}{L_S}} \cos[\theta(t) + \sigma] \right) dt, \quad (3.50)$$

where U_e is the energy loss in time, t, and V_B is barrier voltage drop. Equation (3.50) can be solved by transforming the variable of integration from t to θ using the transformation for dt to $d\theta$ shown in the development of equation (3.39) and by changing the limits of integration using the final equation relating $\theta(t)$ and t in equation (3.39). This technique is demonstrated in section 3.1.1.

3.2 Analysis of the Dual Flying Capacitor (DFC) Bridge

The DFC circuit (Figure 2.14) was introduced in section 2.3.2. As was mentioned in that section, the DFC circuit operation can be viewed as two SFC circuits operating independently. When the two component SFC converters are operated at the same frequency, an arbitrary phase relationship can exist between them. Modulation of this phase does not affect the average rate of energy flow between the coils L_S and L_L. However, this operating phase can change the coil voltage waveforms and thus the frequency spectrum seen by the coils. The waveforms shown in Figure 2.16 are for an operating phase difference of 180° between the SFC converters. In other words, the two switching converters are half a period, $T/2$, out of step. This produces evenly spaced switching intervals and coil voltages which contain only high-order harmonics. The same maximum positive power may be produced by operating the SFC converters at $\pm 90°$ or $\pm T/4$ phase difference, as shown in Figures 3.4 and 3.5. Visual inspection of the v_S and v_L waveforms in these figures reveals that low-order harmonics of large magnitude are present, when compared to Figure 2.16. This can also be verified by a fast Fourier transform (FFT) analysis of the waveforms on a digital computer.

One possible disadvantage of the presence of low-order voltage harmonics is that the coil is more capable of responding to it both in the form of current ripple and possible mechanical vibration. To eliminate low harmonic content in the coil voltages, the two SFC converters should be phased so that all of the DFC switching intervals are equal. Note that the waveforms of v_{C1} and v_{C2} in Figure 2.16 are continuously changing with i_S and i_L. However, equal interval switching is always possible if the operating phase also evolves with i_S and i_L. Figures 3.6, 3.7, and 3.8 show the DFC equal switching interval voltage waveforms at three different instants during the energy transfer.

A special switching pattern for a DFC circuit is the zero operating phase. In this degenerate case, the DFC behaves as one SFC circuit. The result, shown in Figure 3.7, is the worst of possibilities. Not only are high-magnitude, low-harmonic coil voltages present, but the peak system voltages are the highest for the same average power as in Figure 2.16.

3.2.1. Voltages, Currents, and Power

The time periods t_1 and t_3 in Figure 3.8 are the periods in which capacitors C_2 and C_1, respectively, are being charged from 0 to V_0 volts by the source current, $i_S(t) = I_S$ (I_S is considered constant over the switching cycle). Therefore, t_1 and t_3 are numerically given by t_c in equation (3.29). The period t_2 is the time taken to discharge C_1 to zero from its value at the end of t_1. The voltage of C_1 starts from V_0 at

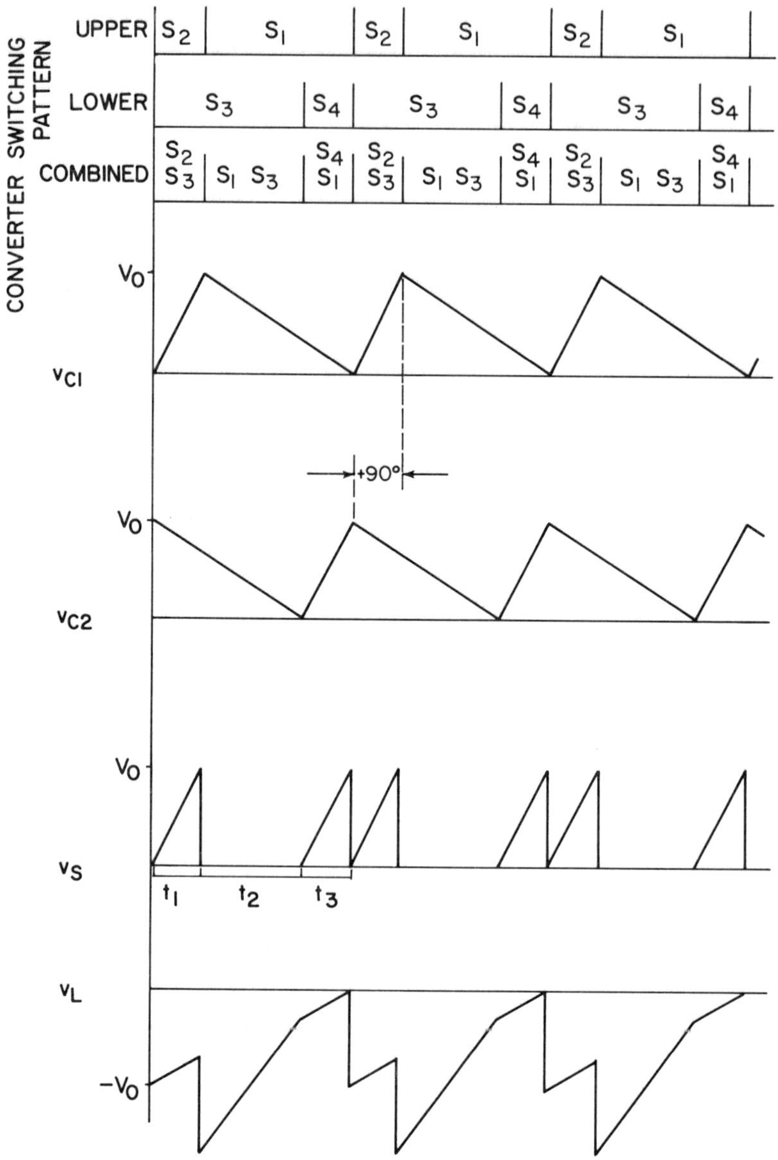

FIG. 3.4. Voltage waveforms of dual flying capacitor (DFC) circuit with an operating phase of $+90°$ and when $i_S > i_L$.

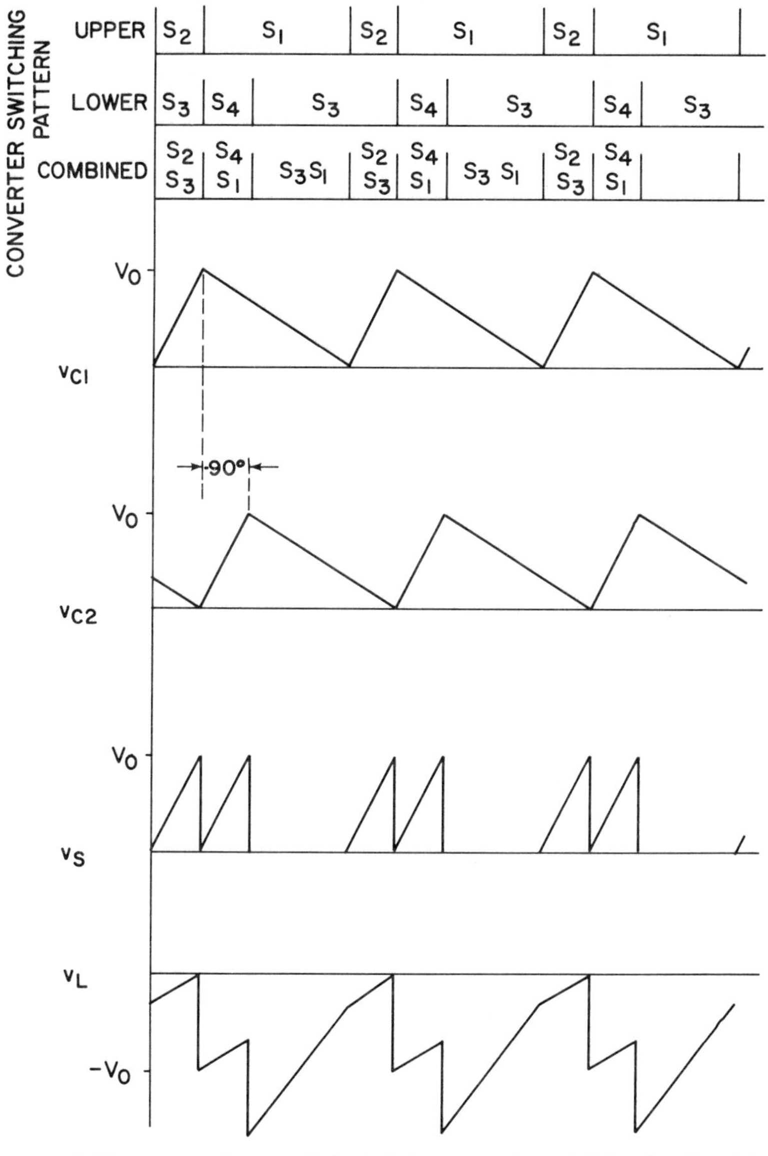

FIG. 3.5. Voltage waveforms of dual flying capacitor (DFC) circuit with an operating phase of $-90°$ and when $i_S > i_L$.

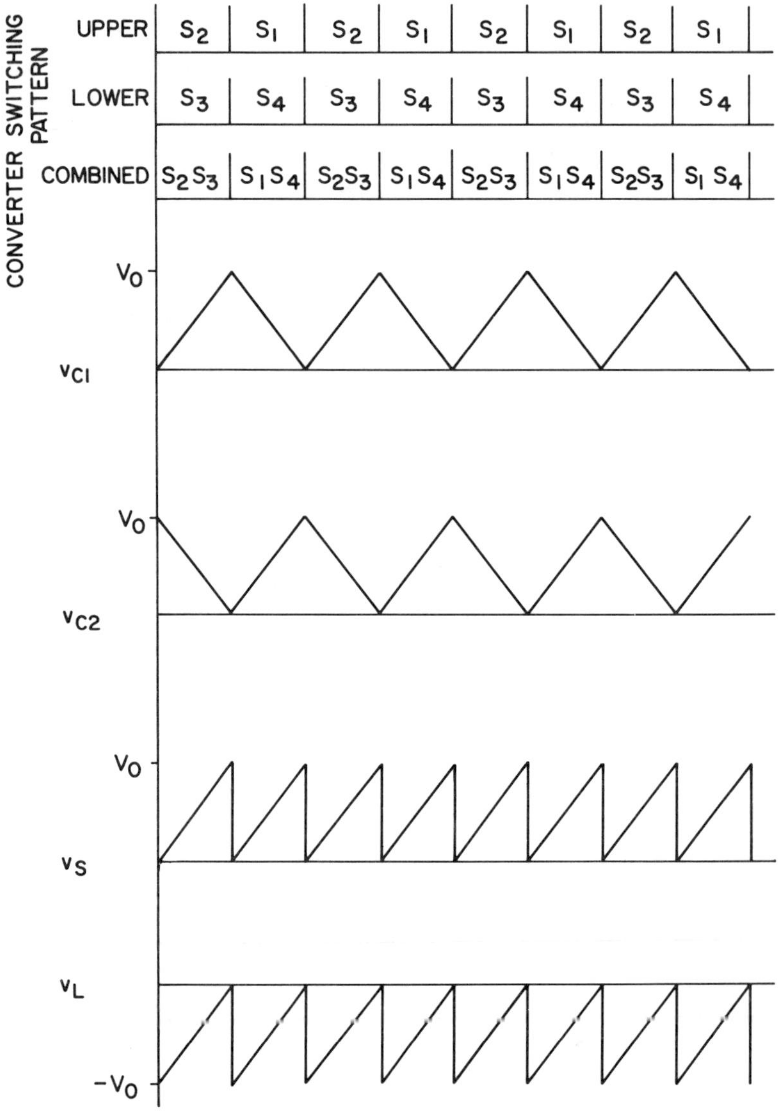

FIG. 3.6. Voltage waveforms of dual flying capacitor (DFC) circuit when $i_S = i_L$.

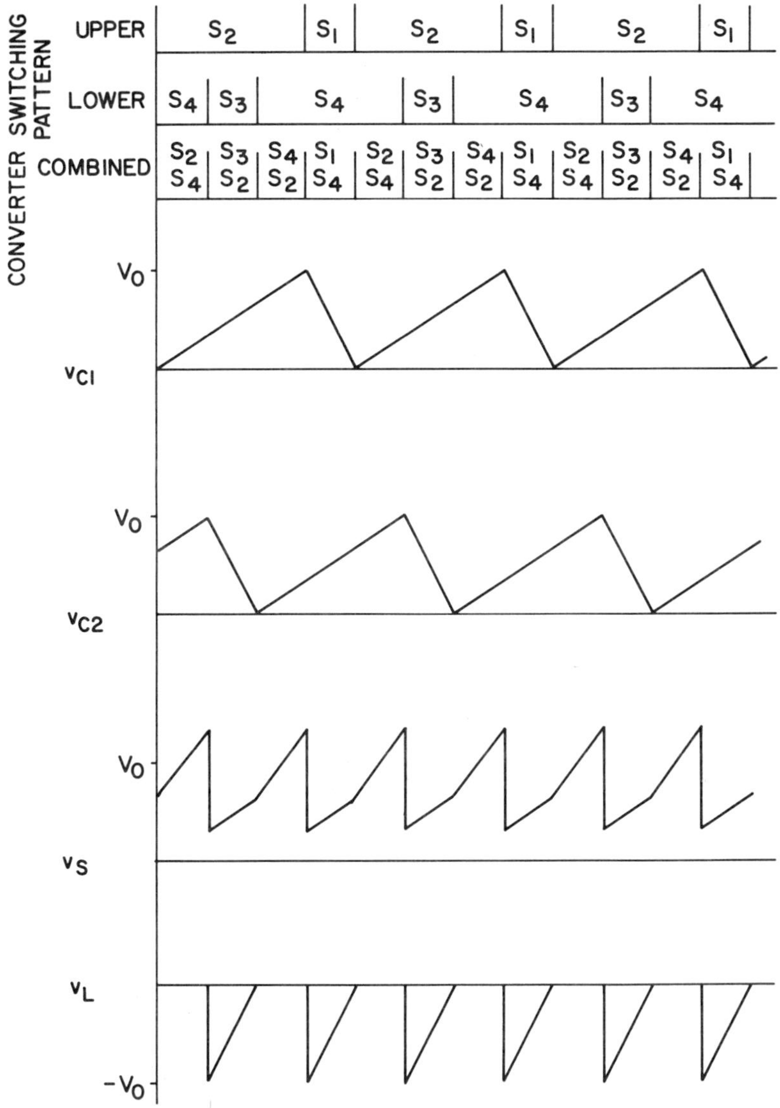

FIG. 3.7. Voltage waveforms for dual flying capacitors (DFC) circuit when $i_S < i_L$.

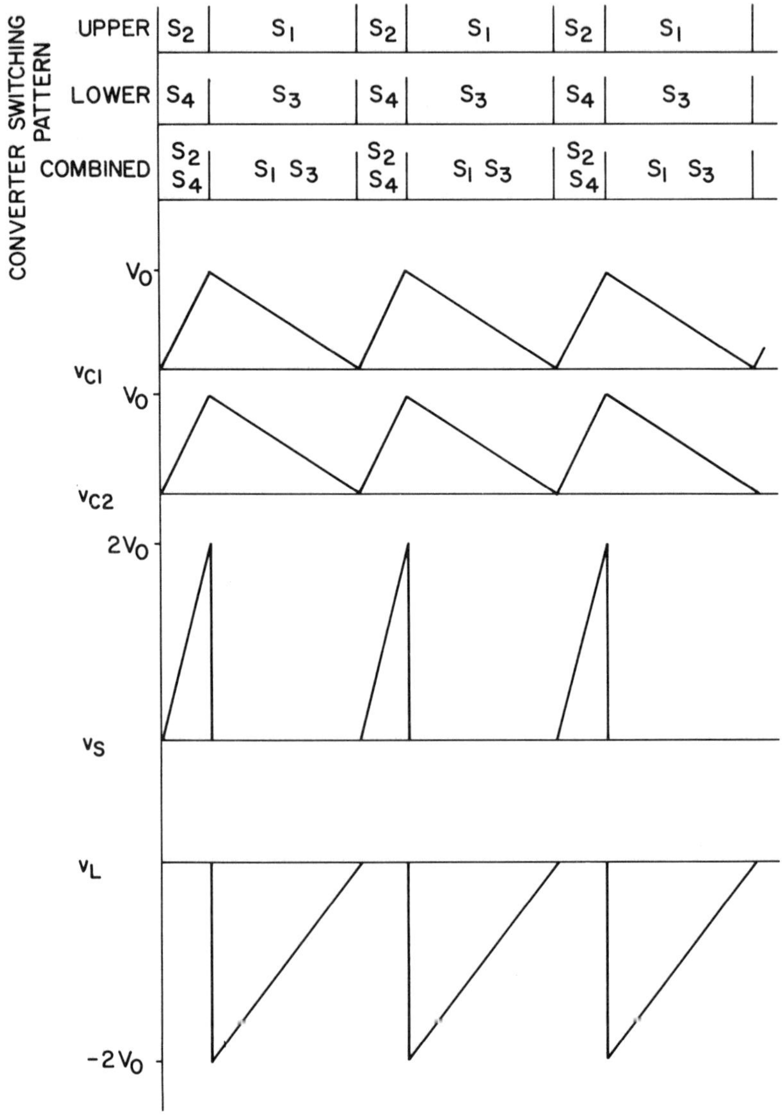

FIG. 3.8. Voltage waveforms for dual flying capacitor (DFC) circuit with an operating phase of 0° and when $i_S > i_L$.

$t = 0$. The voltage on C_1 drops by the amount described in equation (3.51),

$$v_{C1}(t) = V_0 - \frac{I_L t}{C}. \tag{3.51}$$

The total discharge time of C_1 from V_0 to 0 is given by the value of t_d in equation (3.28). The time increment, t_2, is therefore

$$t_2 = t_d - t_c. \tag{3.52}$$

The total switching period, T, is the sum of t_1, t_2, and t_3; i.e.,

$$T = t_c + (t_d - t_c) + t_c;$$

$$T = t_c + t_d \tag{3.53}$$

The average load coil voltage is,

$$\langle v_L \rangle = \frac{1}{T}\int_0^T v_L(t)dt;$$

$$\langle v_L \rangle = -\frac{1}{T}\Bigg[\int_0^{t_c}\left(V_0 - \frac{I_L t}{C}\right)dt + \int_{t_c}^{t_d}\left(2V_0 + \frac{I_L T_c}{C} - \frac{2I_L t}{C}\right)dt + \int_{t_d}^{T}\left(\frac{I_L T}{C} - \frac{I_L t}{C}\right)dt\Bigg];$$

$$\langle v_L \rangle = -V_0\left(\frac{t_d}{T}\right);$$

$$\langle v_L \rangle = -V_0\left(\frac{V_0 C/I_L}{V_0 C/I_L + V_0 C/I_S}\right);$$

$$\langle v_L \rangle = -V_0\left(\frac{V_0 C/I_L}{V_0 C/I_L + V_0 C/I_S}\right);$$

$$\langle v_L \rangle = -V_0\left(\frac{I_S}{I_L + I_S}\right). \tag{3.54}$$

Likewise, the average storage coil voltage is

$$\langle v_S \rangle = V_0 \left(\frac{I_L}{I_L + I_S} \right) \tag{3.55}$$

The energy conservation principle applies to energy transfer with DFC bridges leading to equations (3.33) and (3.34) for $i_S(t)$, $i_L(t)$, and a function of $\theta(t)$, as in the case of the SFC bridge. Equations (3.37) and (3.38) are equally valid for the DFC bridge. The formulation in equation (3.39) is also valid; however, the time-averaged values of coil voltages $\langle v_L \rangle$ and $\langle v_S \rangle$ are exactly twice as large for the DFC bridge as they are for the SFC bridge for a given value of V_0. Thus, $\theta(t)$ is

$$\begin{aligned}\theta(t) &= -\sigma - \beta \\ &\quad + \arccos\left[\frac{-V_0 t}{I_x L_S \sqrt{1 + L_L/L_S}} + \cos(\sigma + \beta) \right],\end{aligned} \tag{3.56}$$

or

$$\begin{aligned}&\cos[\theta(t) + \sigma + \beta] \\ &\quad = -V_0 t / I_x L_S \sqrt{1 + L_L/L_S} + \cos(\sigma + \beta),\end{aligned} \tag{3.57}$$

where β is defined in equation (3.39) and σ is defined in Equation (3.36).

The average power flow between the two coils is:

$$\langle p \rangle = V_0 I_x \frac{\sqrt{\dfrac{L_S}{L_L}} \cos(\theta + \sigma) \sin(\theta + \sigma)}{\sqrt{\dfrac{L_S}{L_L}} \sin(\theta + \sigma) + \cos(\theta + \sigma)}, \tag{3.58}$$

or numerically twice the average power flow for the SFC bridge given in equation (3.42). Similarly, the numerical value for the maximum power flow is $V_0 I_0 \sqrt{2}/4$, twice that of the SFC bridge.

3.2.2 Active Control of the Circuit by Coil Power Reversal

In this section we will consider the control of the time-averaged value of the coil voltages by allowing the capacitor voltages to go to more negative values than what is required simply for current commutation purposes. This technique follows the same logic as for an SFC bridge. In consecutive converter subperiods, each capacitor accepts and then returns energy to a coil. Thus, power reversal takes place between a given coil and capacitor during each converter cycle. The direction of net energy flow depends on the excess of the energy delivered or accepted by the coil. Sample waveforms for this control mode are shown in Figures 3.9 and 3.10.

Circuit functions in the periods t_1, t_2, and t_3 are basically the same; i.e., C_2 is charged by I_S during t_1; C_1 is charged by I_L during t_3; and L_S is shorted during t_2. The waveforms in Figures 3.9 and 3.10 are generated by allowing the time increments to increase by amounts proportional to t_c and t_d yet maintaining $\pm V_0$ as the maximum capacitor voltage; i.e.,

$$t_1 = t_3 = t_c + kt_c,$$

$$t_2 = t_d - t_c + k(t_d - t_c) \tag{3.59}$$

where $0 \leq k \leq 1$ and t_c and t_d are as defined in section 3.1.2. The switching period, T, is equal to $(t_d + t_c)(1 + k)$. At the end of period t_1 the capacitor voltage on C_2 is charged to V_0 by the storage coil current, i.e.,

$$v_{C2}(t) = V_0. \tag{3.60}$$

The voltage on C_1 during the same time increment is discharged by the load coil current from V_0; i.e.,

$$v_{C1}(t) = V_0 - \frac{(1 + k)t_c I_L}{C};$$

$$v_{C1}(t) = V_0 - (1 + k)\frac{I_L}{I_S}V_0. \tag{3.61}$$

During the time increment, t_2, C_1 and C_2 are both discharged by the load coil current. The voltage on C_2 starts the discharge from the value of V_0 at the end of the t_1 time increment and reaches the following value at the end of the t_2 time increment:

$$v_{C2}(t) = V_0 - \frac{(1 + k)(t_d - t_c)I_L}{C};$$

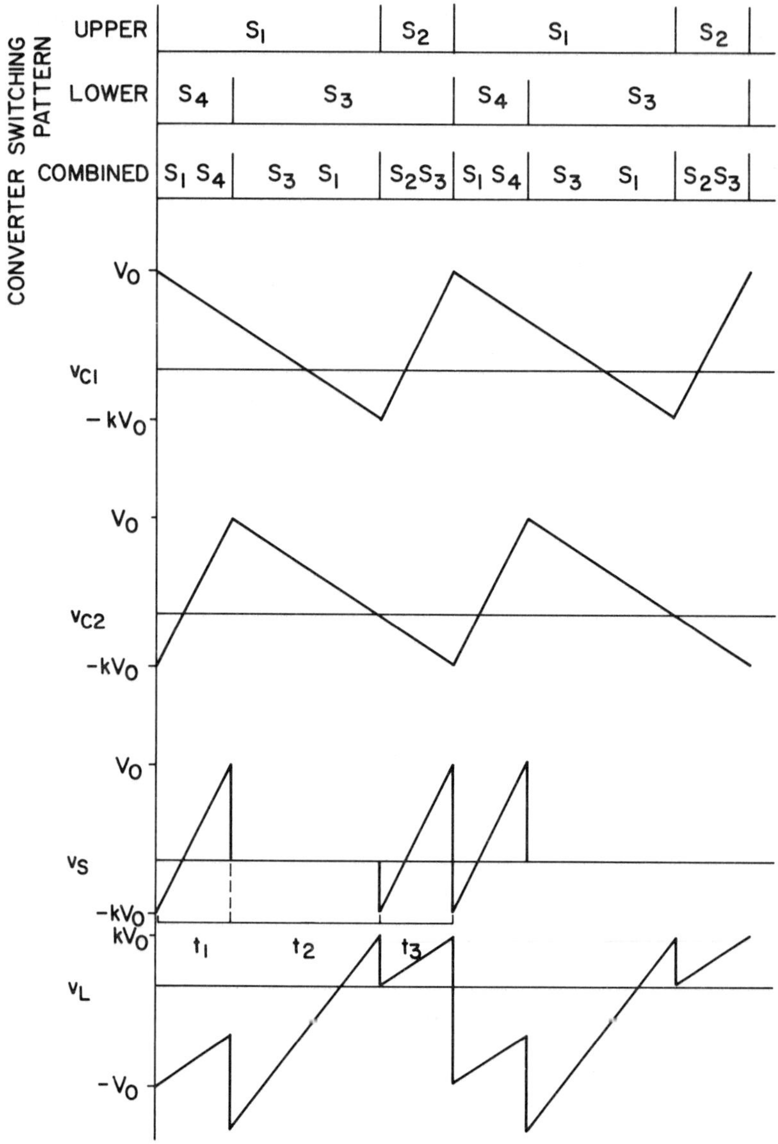

FIG. 3.9. Voltage waveforms of dual flying capacitor (DFC) with oscillating-power active control, $0 \leq k \leq 1$.

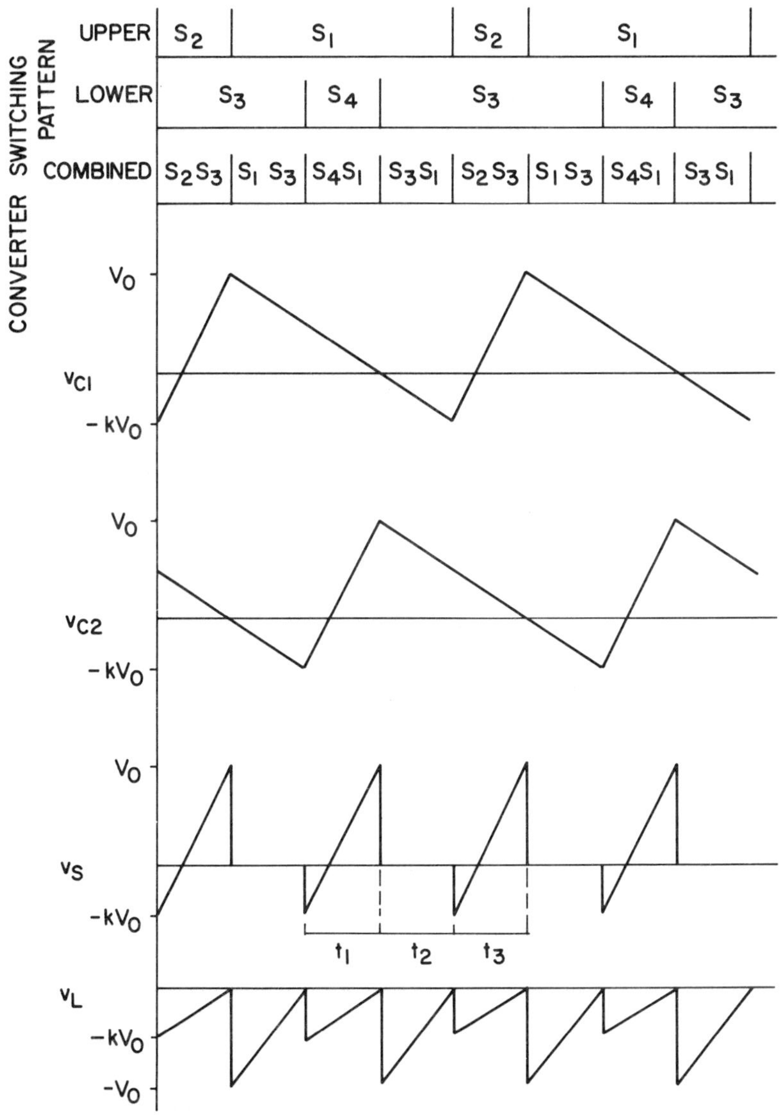

FIG. 3.10. Voltage waveforms of dual flying capacitor (DFC) with oscillating-power active control, equal switching intervals and $0 \leqslant k \leqslant 1$.

$$v_{C2}(t) = V_0 - (1 + k)V_0 + (1 + k)\frac{I_L}{I_S}V_0;$$

$$v_{C2}(t) = kV_0 + (1 + k)\frac{I_L}{I_S}V_0. \tag{3.62}$$

The voltage $v_{C1}(t)$ was initially at V_0 at the start of t_1 and was reverse charged by the end of the time increment t_2. Consequently,

$$v_{C1}(t_1 + t_2) = V_0 - (1 + k)\frac{I_L}{I_S}V_0 - \frac{(1 + k)(t_d - t_c)I_L}{C};$$

$$v_{C1}(t_1 + t_2) = V_0 - (1 + k)\frac{I_L}{I_S}V_0 - (1 + k)V_0 + (1 + k)\frac{I_L}{I_S}V_0;$$

$$v_{C1}(t_1 + t_2) = -kV_0. \tag{3.63}$$

During the time increments t_3, C_1 is charged by I_S from a voltage of $-kV_0$ to V_0 and C_2 continues to be discharged by I_L. Therefore, at the end of the t_3 period,

$$v_{C1}(t) = V_0. \tag{3.64}$$

The voltage on C_2 starts the t_3 time increment with the value given by equation (3.62) and is discharged at a rate proportional to I_L/C; therefore, at the end of the t_3 period,

$$v_{C2}(t) = -kV_0 + (1 + k)\frac{I_L}{I_S}V_0 \quad \frac{(1 + k)t_cI_L}{C};$$

$$v_{C2}(t) = -kV_0 + (1 + k)\frac{I_L}{I_S}V_0 - (1 + k)\frac{I_L}{I_S}V_0;$$

$$v_{C2}(t) = -kV_0. \tag{3.65}$$

Thus, the two capacitors are charged and discharged between the values of $-kV_0$ and V_0. The resulting storage coil voltage is shown in Figure 3.9. The average value of the storage coil voltage is

$$\langle v_S \rangle = \frac{V_0 t_c - kV_0(kt_c)}{T};$$

$$\langle v_S \rangle = (1 - k^2)V_0\left(\frac{t_c}{(1 + k)(t_c + t_d)}\right);$$

$$\langle v_S \rangle = (1 - k)V_0\left(\frac{I_L}{I_L + I_S}\right). \tag{3.66}$$

Since power flow is constant between the two coils, the average load coil voltage is

$$\langle v_L \rangle = (1 - k)V_0\left(\frac{I_S}{I_L + I_S}\right). \tag{3.67}$$

Equations (3.66) and (3.67) show that the average coil current can be controlled by the control variable, k. The variable k is modulated by varying the duration of conduction patterns on each SFC component of the DFC converter. For example, to increase the average power from the level shown in Figure 3.10, the peak negative voltage on the capacitors must be reduced as shown in Figure 3.11. This is accomplished by shortening the S_1 conduction to reduce the negative peak on v_{C1} and by shortening S_2 conduction to keep the positive peak on v_{C1} at the V_0 rating. Similar steps are taken for S_3 and S_4 to adjust v_{C2}.

Note that the component SFC converters will now be operating at a different frequency. Therefore, the same timing difference between the upper and lower SFC will translate to a new phase relationship after the change of k. This change of operating phase will change the intended frequency spectrum and peak voltages seen by the coils. In Figure 3.11, the operating phase is seen to have changed from 270° to −90° after k was reduced to zero to maximize power.

To compensate for this inadvertent change of phase, a phase-shifting switching transient is imposed on one of the SFC converters. For example, to bring the two SFC converters of Figure 3.11 to a desired 180° phase relationship, the consecutive conduction times of S_4 and S_3 should be shortened appropriately, as shown in Figure 3.12. This phase-shifting switching transient could be merged with the k changing transient by the switching controller.

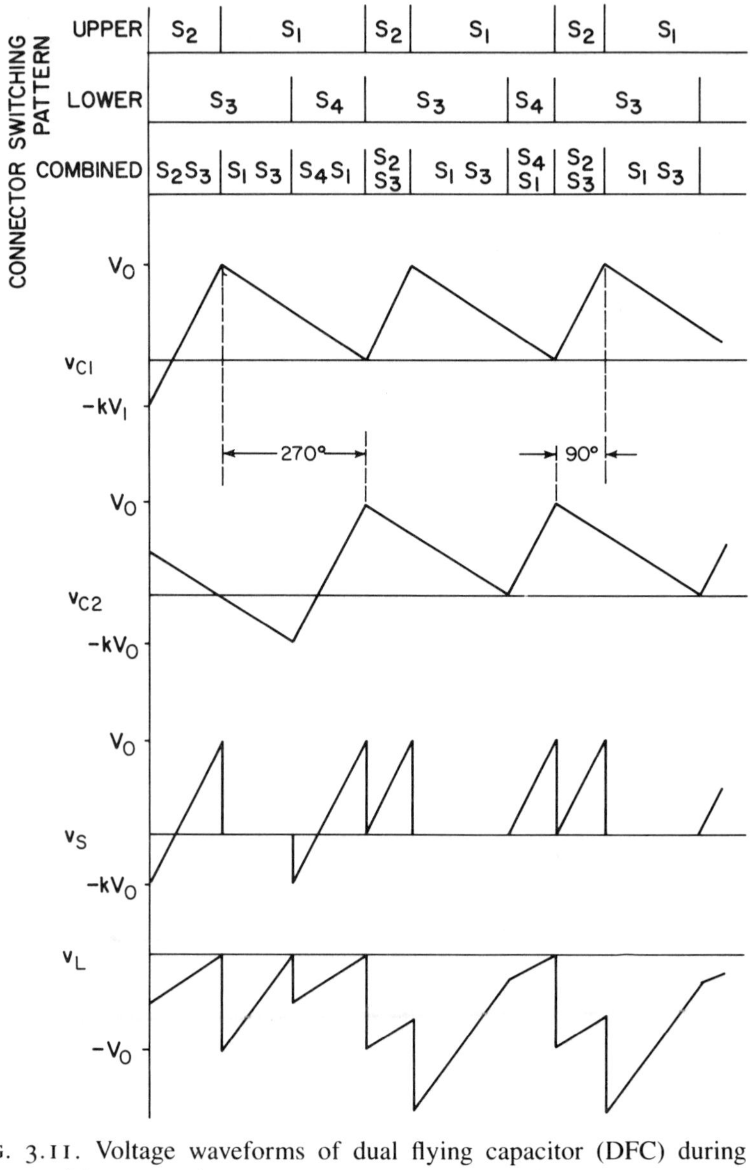

FIG. 3.11. Voltage waveforms of dual flying capacitor (DFC) during the change of k to zero when $i_S > i_L$.

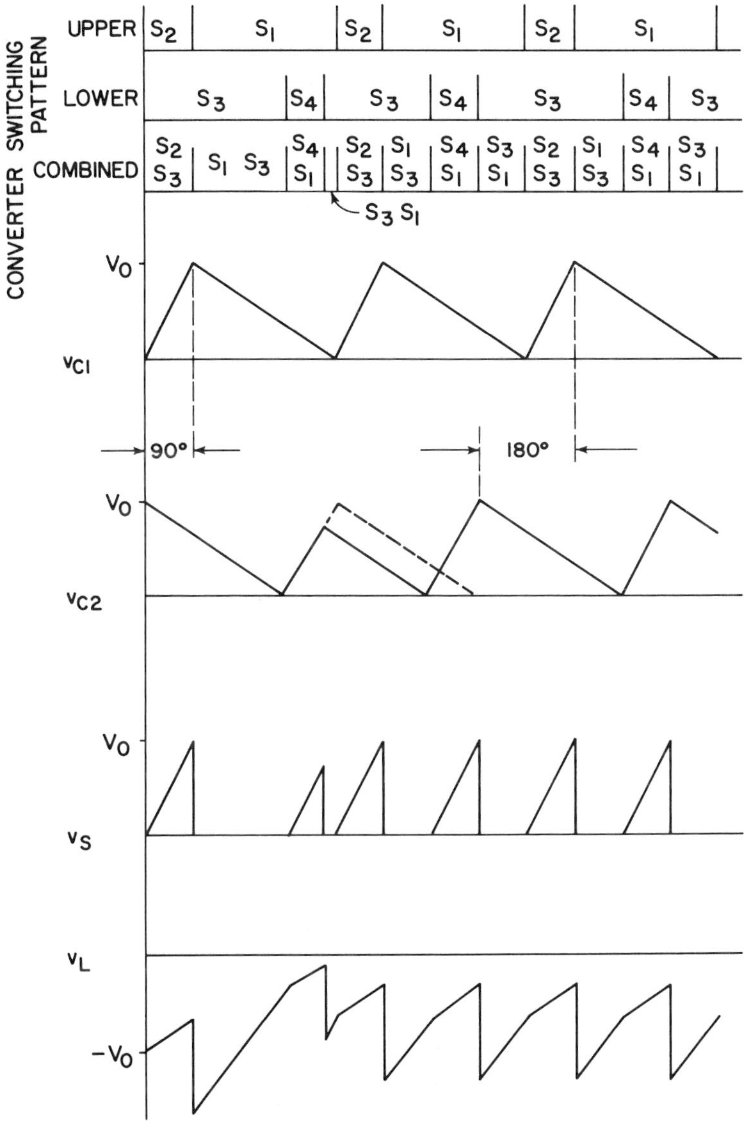

FIG. 3.12. Voltage waveforms of the dual flying capacitor (DFC) during phase shifting 90° → 180° when $i_S > i_L$.

Two important constraints must be observed whenever a change of switching times is contemplated. Overvoltages beyond the V_0 rating on the capacitors must be avoided or voltage breakdown failures may result in the system. Proper positive and negative minimum peak voltages must be maintained on the capacitors for the commutations in each SFC. The above constraints will dictate the pattern of the switching transients for any change in the operating state, such as the magnitude and sign of system power or the operating phase.

3.2.3 Component Ratings

The capacitive energy storage, U_C, in a DFC bridge is equal to twice the stored energy per capacitor, or

$$U_C = 2\left(\frac{1}{2}CV_0^2\right);$$

$$U_C = \frac{CI_0^2t_c^2}{C^2};$$

$$U_C = \frac{I_0^2t_c^2}{C}. \tag{3.68}$$

For an equivalent switching time and initial storage coil current, the installed capacitive energy storage would be twice as large as the SFC bridge, except, of course, that the defining conditions are determined by the load coil control voltage. In the DFC bridge, the time-average value of load coil voltage has twice as large a value for a given capacitor voltage than is generated in an SFC bridge. Therefore, for a given control coil voltage and value of I_0, the value of capacitance must be twice as large in the DFC bridge to gain only half the voltage in the same time increment. Thus, total installed capacitive energies are the same in the two bridges.

The ratio of peak voltage to average voltage is

$$\frac{V_0}{\langle v_L\rangle} = -\left(\frac{I_L + I_S}{I_S}\right), \tag{3.69}$$

or, in other words, one-half of the equivalent value, equation (3.46) for the SFC bridge. Of course, the DFC bridge has four switching elements, whereas the SFC has only two, so the total number of thyristors is equal in the two bridges.

The rms current ratings of the devices are of the same algebraic values as presented in equations (3.47), (3.48), and (3.49).

3.2.4 Efficiency of Energy Transfer

A single switch element in the DFC bridge always carries the sum of the two coil currents during its conduction period, as was the case for the SFC bridge. The coil currents in the DFC bridge must always pass through two switch elements in the loop containing one of the capacitors and the coil. Therefore, the equation for efficiency of the DFC is twice the algebraic value of equation (3.50).

$$U_e = \int_0^t \frac{2V_0 V_B}{V_T} I_x \sqrt{\frac{L_S}{L_L}} \times \left(\sin(\theta + \sigma) + \sqrt{\frac{L_L}{L_S}} \cos(\theta + \sigma) \right). \qquad (3.70)$$

The numerical value of V_0 for the DFC bridge is one-half as large as the value for the SFC bridge if the same time-averaged value of coil voltage is used for both bridges. Therefore, the efficiency of energy transfer is the same for either type of bridge for the same time-averaged value of coil voltage.

CHAPTER 4

Uniform Switching Energy Transfer Circuits: Inductor-Converter Bridges

The multiphase inductor-converter bridge (ICB) was introduced by H. A. Peterson, N. Mohan, W. C. Young and R. W. Boom of the Superconductive Energy Storage Group of the University of Wisconsin. Unlike the SFC, S^4LCL, DFC, and other circuits which have time-varying switching intervals for a constant time-averaged value of coil voltage, the ICB circuits operate with a clock-driven sequence of gate pulses which have constant switching intervals for a constant time-averaged value of coil voltage. The ICB circuits, therefore, belong to a new group of converters which are uniformly switching during the energy transfer.

In this chapter the fundamentals of operation and control of the important members of the ICB group will be presented. Chapter 5 contains the more rigorous analysis of these converters.

4.1 The One-Phase ICB Circuit

The one-phase ICB was proposed by Ehsani and Kustom at Texas A&M University and Argonne National Laboratory. A schematic circuit diagram of the one-phase ICB is shown in Figure 4.1

4.1.1 Circuit Operation

The storage coil, L_S, and load coil, L_L, are each connected to a full-wave one-phase converter as shown. The ac lines of the converters are connected in parallel with the capacitor C. The switching sequence on the storage converter is S_{S1} S_{S4}, S_{S2} S_{S3}, S_{S1} S_{S4}, etc. The same switching sequence and frequency are used on the load converter. However, corresponding switching events on the storage and load converters need not be synchronized. Leading timing (leading phase), less than half a cycle, of the load converter relative to the storage converter will cause a net energy transfer from storage to load coil. Similarly, lagging load converter timing (lagging phase) will cause net energy removal from the load coil. The capacitor on the ac side of converters serves two functions. It temporarily stores the energy which is transferred from one coil to the other in each converter cycle. It also supplies the reverse voltages required for commutation of current between S_1S_4 and S_2S_3 switches on each converter.

Although the above operation is generally similar to that of other ICB circuits described later, an instantaneous double short circuit occurs across each inductor during commutation, twice a cycle, which does not happen in the other ICB systems. For example, S_{S1}, S_{S2}, S_{S3}, and S_{S4} switches are in conduction and a double-path short circuit exists on L_S. A positive voltage on the capacitor C is required to commutate S_{S1} and S_{S4} off while maintaining conduction in S_{S2} and S_{S3}. However, this double-short conduction pattern is not inherently damaging, because short circuit across a large inductor maintains a constant current. Switch S_{S1} is turned off by the capacitor voltage in the commutation loop consisting of S_{S1}, S_{S2}, and C. With the capacitor voltage positive, as shown in Figure 4.1, reverse S_{S1} current will flow in the mentioned loop. As stated in chapter 1, when this commutation current equals the forward current, i_S, carried by S_{S1}, this SCR will open as in a reverse-biased diode. Note that the inductance and resistance of the commutation loop S_{S1}, S_{S2}, and C is very small by design. Therefore, small capacitor voltages can set up very high amplitude commutation currents which are sufficient for diverting i_S away from S_{S1}. A similar commutation current occurs in the loop S_{S4}, S_{S3}, and C which turns off S_{S4}. The duration of this commutation, called overlap, is very short relative to the converter cycle.

It is noteworthy that significant electromagnetic noise, common in SCR circuits, is radiated from these high-current short-pulse loops. Four such double commutations take place in each cycle of the one-

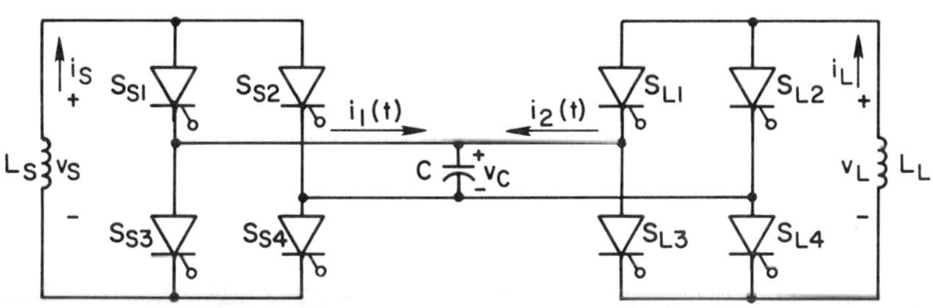

FIG. 4.1. Circuit diagram of the one-phase inductor-converter bridge (ICB).

phase ICB. It is evident from the above discussion that correct capacitor voltage polarity must be present for successful commutation of one pair of SCRs to the next in each converter. The limitations imposed by this voltage polarity will be considered later. However, it is a simple matter to verify on the diagram of Figure 4.1 that turning off SCRs $S_{S1}S_{S4}$ and $S_{L1}S_{L4}$ requires positive capacitor voltage. Therefore, an initial voltage; e.g., $v_C < 0$ is placed on the capacitor, and the converters are always started with the same SCR pairs; e.g., $S_{S1}S_{S4}$ and $S_{L1}S_{L4}$.

The circuit usually begins operation with the storage inductor fully charged and the load inductor uncharged. The converter switching frequency is chosen such that the total storage energy is transferred to the load over several thousand converter cycles. Therefore, each converter cycle transfers a very small fraction of the system energy. Thus, the coil currents are almost unchanged during a single converter cycle. During the first few cycles of operation the capacitor current is almost entirely from the storage converter, $i_1(t)$ in Figure 4.1. This current waveform, along with the resulting capacitor voltage, is shown in Figure 4.2. The waveforms in this figure are for the case of load converter switching leading the storage converter by a quarter cycle or a phase difference of $\phi = 90°$. The impulse gate currents, i_g, show the timing of each SCR pair on the storage and load converters. Note that for converter operation starting with $S_{S1}S_{S4}$, a negative capacitor precharge voltage, v_{C0}, is present. Referring to Figure 4.2, note that the capacitor voltage oscillates between $+v_{C0}$ and $-v_{C0}$. Balanced capacitor voltage oscillation requires that v_{C0} be precisely equal to peak capacitor voltage calculated below.

$$v_{C0} = \frac{TI_{S0}}{4C} \tag{4.1}$$

where T is the converter period and I_{S0} is the initial storage coil current. However, it will be shown that any capacitor voltage unbalance caused by inaccuracy of precharge voltage will decay to zero. In other words, the circuit can automatically correct for capacitor voltage errors. This notion may provide several benefits to be discussed later. The storage and load coil voltages, v_S and v_L, shown in Figure 4.2 are with respect to the positive voltage convention of the coils in Figure 4.1. Observe that the initial average voltages on the storage and load

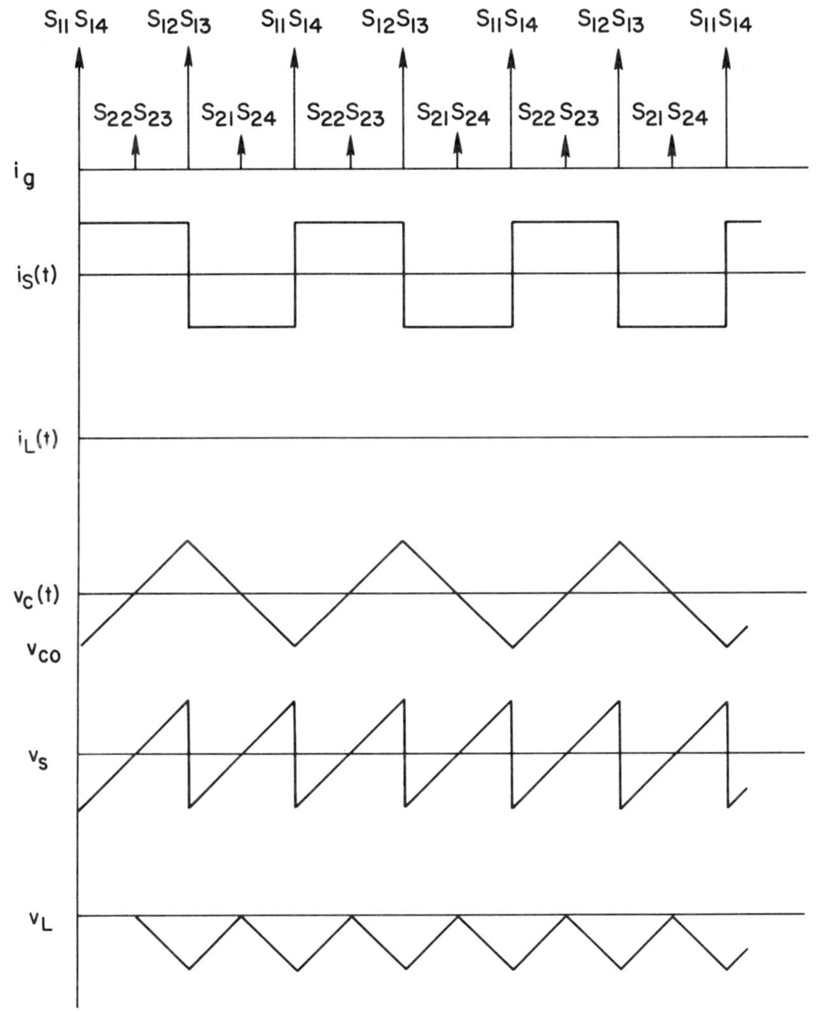

FIG. 4.2. One-phase ICB waveforms during the first few cycles with operating phase $\phi = 90°$.

coils are zero and negative, respectively. This will be shown more clearly in the theoretical analysis of the system in the next chapter. It is also interesting to note that $S_{S1}S_{S4}$ and $S_{S2}S_{S3}$ commutations have ample capacitor voltages of the right polarity. However, $S_{L1}S_{L4}$ and $S_{L2}S_{L3}$ have to commutate when the capacitor voltage is almost zero. Therefore, $i_S = i_{S0}$ and $i_L = 0$ is a threshold state for successful com-

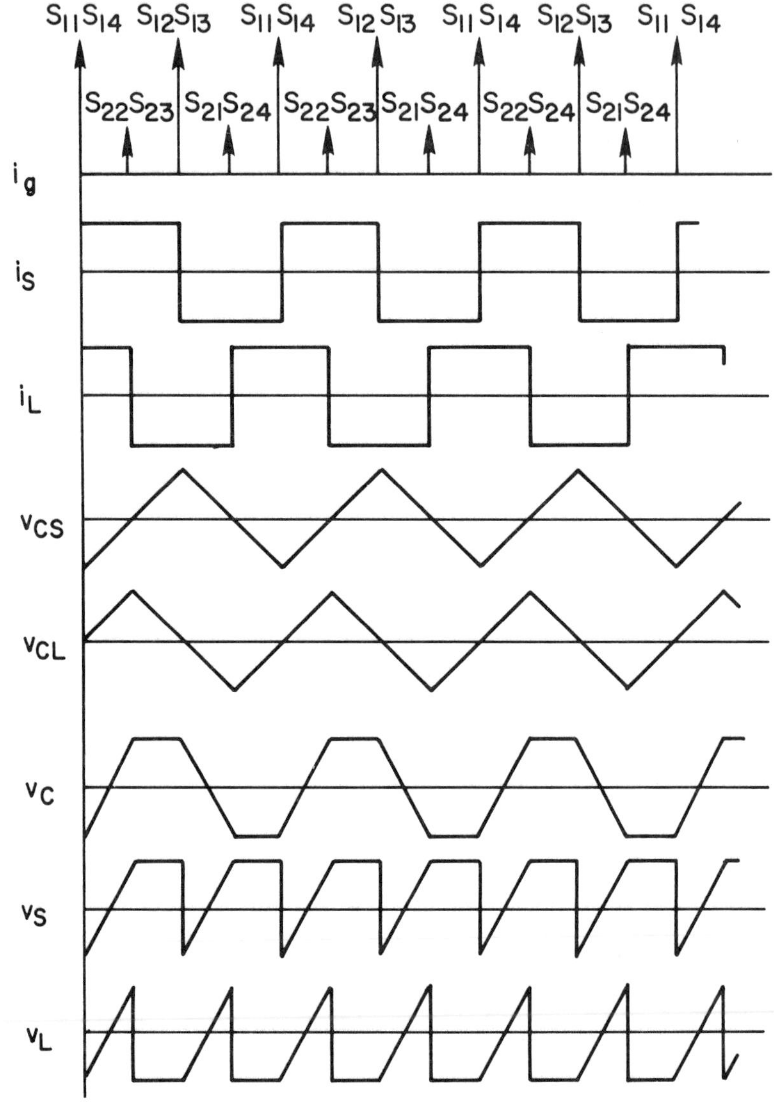

FIG. 4.3. One-phase ICB waveforms at $i_L = i_S$ and $\phi = 90°$.

mutation at $\phi = 90°$. It will be shown analytically that the commutation voltage will progressively improve as i_L builds up.

Figure 4.3 shows the circuit waveforms when $i_L = i_S$. For a system with equal storage and load inductors ($L_S = L_L$) this is the point of equal energy in the coils. The capacitor voltage component resulting from storage converter current i_1, is shown as v_{CS}, while the component due to the load converter current, i_2, is shown as v_{CL}. The actual capacitor voltage, v_C, is the sum of these two components. The coil voltages are derived from v_C and by observing which SCR pairs are conducting on each converter at a given time. It is observed that a net positive voltage is present on L_S and a negative net voltage exists on L_L. These voltages are responsible for extracting net energy from L_S and adding it to L_L. Note, however, that oscillating coil voltages against direct coil currents produce oscillating instantaneous coil powers. Also note that ample capacitor commutation voltages of the correct polarity exist for S_1S_4 and S_2S_3 commutations on both converters. For completeness, the circuit waveforms at the end of energy transfer, where $i_S = 0$ and $i_L = i_{L\max}$, are shown in Figure 4.4. It is seen that capacitor voltage, v_C, is entirely determined by the load converter current, i_2. The coil voltages show a high net voltage on L_S and a zero net voltage on L_L. Circuit waveforms at numerous other energy states, i_L/i_S and phase angles, ϕ, may similarly be drawn.

4.1.2 Circuit Control

The direction and rate of energy flow between the two coils are determined by the relative timing or phase difference between the storage and load converters of a one-phase ICB. For a net energy transfer from L_S to L_L, the load converter must lead the storage converter and vice versa. It will be shown quantitatively in the next chapter that the lead time or operating phase of the load converter determines the rate of energy flow. For maximum power, the operating phase turns out to be 90°.

The frequency or rate of switching of the one-phase ICB also controls the converter power. However, the direction of energy flow cannot be controlled by the frequency. Therefore, frequency of the ICB is usually modulated as a supplementary control variable to trim and optimize the system operation. The explicit mathematical relationship of the frequency to power is derived in the next chapter.

For real-time control of power, the operating phase and frequency of the ICB are modulated by the controller. This control requires a series of specific converter switching transients to maintain proper commutation voltage and safe voltage levels on the capacitor. These transient switching sequences are discussed in chapter 5.

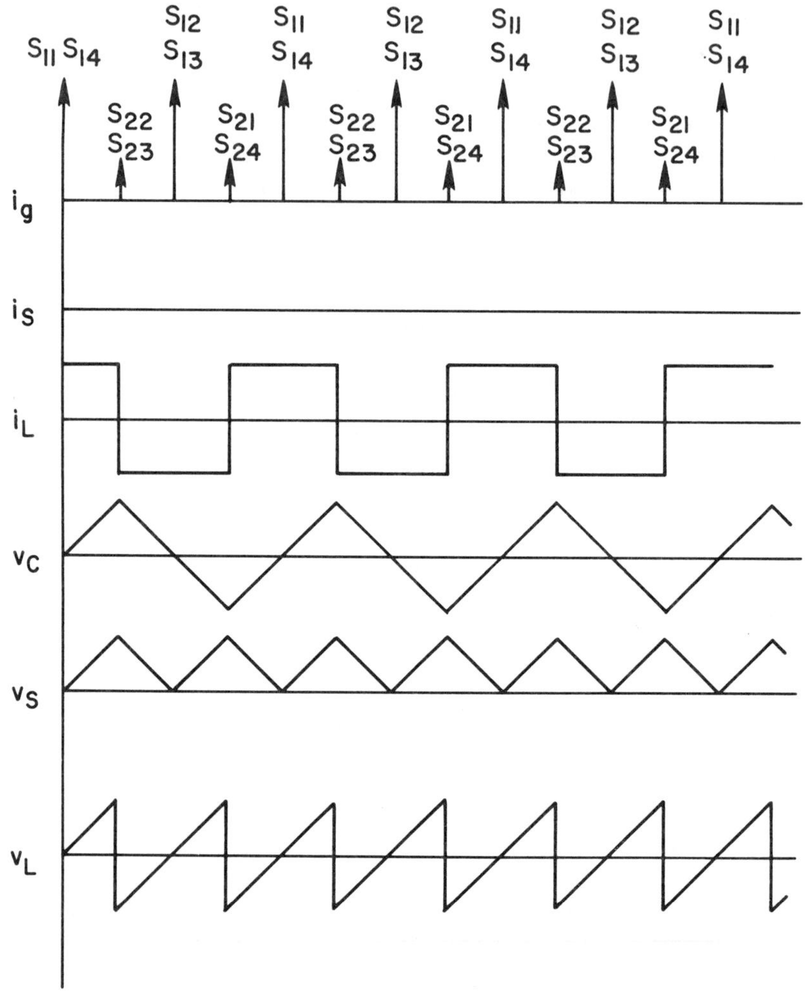

FIG. 4.4. One-phase ICD waveforms at $i_S = 0$, $i_L = i_{L\,\max}$, and $\phi = 90°$.

4.1.3 Component Rating

The installed value of capacitive energy storage, U_C, for the one-phase ICB is

$$U_C = \frac{1}{2} C V_m^2, \tag{4.2}$$

where V_m is the maximum possible capacitor voltage during the energy transfer.

It can be verified by inspection of Figure 4.3 that the highest capacitor voltage is generated at zero phase angle. At this operating phase, converter currents from both converters add up to produce the capacitor voltage. Therefore,

$$V_C = \frac{t_{sw}}{C}(I_S + I_L), \tag{4.3}$$

where t_{sw} is the duration of switch conduction. If I_L is assumed to be equal to kI_S, where k is a constant between 0 and 1, then

$$V_C = \frac{I_S t_{sw}}{C}(1 + k). \tag{4.4}$$

The currents I_S and I_L are related to each other through equation (3.32). If one assumes that the initial system energy, U_0, is equal to $1/2(L_S I_0^2)$, equation (4.4) becomes

$$V_C = \frac{I_0 t_{sw}}{C}\left(\frac{1 + k}{\sqrt{1 + k^2 L_L/L_S}}\right). \tag{4.5}$$

The maximum value of capacitor voltage occurs when k is equal to L_S/L_L. When L_S is equal to L_L, the value of installed capacitive energy is

$$U_C = \frac{1}{2}CV_m^2,$$

$$U_C = \frac{1}{2}C\left(\frac{2I_0^2 t_{sw}^2}{C^2}\right),$$

$$U_C = \frac{I_0^2 t_{sw}^2}{C}. \tag{4.6}$$

The maximum forward and reverse holding voltages of each switch element are equal to the maximum capacitor voltage. This can be verified on Figure 4.1 when two switches are on and the other two are off.

The highest current a switch element must carry is equal to the highest current in the coil corresponding to the side in which the switch is located. Such switch element conducts for one-half of the convert cycle. Therefore, the rms current rating per switch element is

$$I_{\text{rms}} = \sqrt{\frac{1}{T}\int_0^{T/2} I_0^2 dt},$$

$$I_{\text{rms}} = I_0/\sqrt{2}. \tag{4.7}$$

4.1.4 Efficiency of Energy Transfer

The current from each coil flows through two switching elements at any given time. If the forward conduction voltage drop of each switching device is V_B, the total conduction drop for each coil current is $2V_B$ times the number of series devices per switching element. The number of switching devices per switch element is V_m/V_T, where V_m is equal to $I_0 t_{sw}\sqrt{(1 + L_S/L_L)/C}$ and V_T is the repetitive holding voltage rating of the switching device.

Therefore, the total conduction losses will be

$$U_e = \int_0^t \frac{2V_m V_B}{V_T} I_x \cos\theta(t) dt,$$

$$+ \int_0^t \frac{2V_m V_B}{V_T} I_x \sqrt{\frac{L_S}{L_L}} \sin\theta(t) dt. \tag{4.8}$$

Equation (4.8) has the same form as equation (3.70).

For total converter losses, the switching and other losses must be added to the above conduction losses. These losses have been discussed in section 3.1.4.

4.2 The Three-Phase ICB Circuit

The three-phase ICB was the original ICB circuit invented by Peterson and colleagues. A schematic circuit diagram of the three-phase ICB is shown in Figure 4.5.

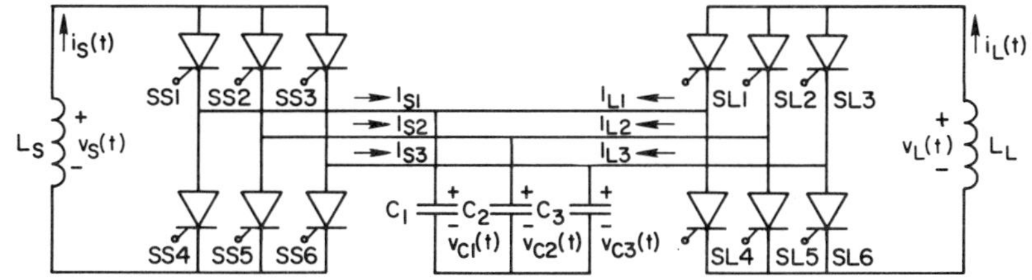

FIG. 4.5. Circuit diagram for three-phase inductor-converter bridge (ICB).

4.2.1 Circuit Behavior

Each three-phase (six-pulse) converter in Figure 4.5 is switched in the normal Graetz bridge sequence. In steady state, this switching occurs in equal intervals of time as follows: SS1SS5 to SS1SS6 to SS2SS6 to SS2SS4 to SS3SS4 to SS3SS5 to SS1SS5, and so on. The load-side bridge follows the same switching sequence and at the same frequency, but not necessarily in synchronism with the storage side.

At the beginning of the energy transfer, all the energy is in L_S. Capacitors C_1, C_2, and C_3 are independently precharged to $-V_0$, 0, and $+V_0$, respectively. The storage side SCRs SS1 and SS5 are gated into conduction. The storage coil current, $i_S(t)$, circulates in the loop consisting of L_S, SS1, C_1, C_2, and SS5, charging C_1 positively and C_2 negatively. Positive convention is defined in Figure 4.5. The voltage of these capacitors will tend to change sinusoidally. However, if the conduction duration, Δt, in this mode is much shorter than the natural period of the LC loop, the capacitor voltages may be described, to a high accuracy, by the linear terms of their Taylor expansion about the initial voltage,

$$V_{C1} = -V_0 + i_S \Delta t / C, \tag{4.9}$$

$$V_{C2} = 0 - i_S \Delta t / C. \tag{4.10}$$

During this time interval, the voltage on C_3 is unchanged. At the end of this interval, switch SS6 is turned on and i_S is naturally commutated from SS5 to SS6. This commutation condition can be checked by determining the polarity of the voltage on each switch around the loop consisting of SS5, SS6, C_2, and C_3:

$$-V_{SS5} + V_{C2} - V_{C3} + V_{SS6} = 0, \tag{4.11}$$

where positive V_{SS5} and V_{SS6} convention is assumed at the anodes. Substituting $V_{SS5} = 0$ for the conducting switch and the final values of the capacitors in equation (4.11),

$$0 - i_S\Delta t/c - V_0 + V_{SS6} = 0, \tag{4.12}$$

$$V_{SS6} = V_0 + i_S\Delta t/C. \tag{4.13}$$

If the initial voltage on the capacitor, V_0, is chosen properly, SS6 is forced into conduction, $V_{SS6} > 0$. Upon conduction, $V_{SS6} = 0$ and the loop current will reverse-bias SS5:

$$-V_{SS5} - i_S\Delta t/C - V_0 + 0 = 0, \tag{4.14}$$

$$V_{SS5} = -(V_0 + i_S\Delta t/C) < 0. \tag{4.15}$$

If the condition of equation (4.15) holds for longer than the turn-off time of SS5, this SCR will turn off, regaining its open-circuit state, and the SS1SS6 mode of conduction begins. Similar voltage polarity conditions must be met for the transition from any conduction mode to the next. It can be shown that precharging the capacitors to $V_0(\because i_S\Delta t/C > 0) = i_S\Delta t/C$ will insure successful commutation in all six modes of conduction and will produce a balanced set of three-phase capacitor voltages. Figure 4.6 shows the voltage waveforms on the capacitors and the storage coil.

When the load current is not equal to zero, the capacitors are charged from both inductors. However, since negligible inductive energy is extracted during each converter conduction mode, the inductor currents are almost unchanged during the interval. Thus, the capacitors may be assumed to be charged from the superposition of the load and storage converter currents. The above statements are equivalent to approximating the capacitor voltage during the time between switching events by the linear terms of its Taylor expansion about the initial voltage,

$$V_C(t) = V_0 \pm i_S\Delta t'/C \pm i_L\Delta t'/C. \quad 0 \leqslant \Delta t' < \Delta t \tag{4.16}$$

The $\pm$ signs depend on the particular capacitor and conduction modes involved, and $\Delta t'$ is the duration of this combination of modes. Figure 4.7 is an example of a capacitor voltage waveform generated by several intervals of equation (4.16) when $L_S = L_L$, $i_S = 2i_L$, and the load bridge is switched exactly one switching step ahead of the storage; i.e., either $\Delta t' = \Delta t$ or $\Delta t' = 0$. From equation (4.16) it is evident that the commutation conditions for both the storage and load converters will depend on the current magnitude in both the storage and load, i_S and i_L, as well as the relative timing of the converters which deter-

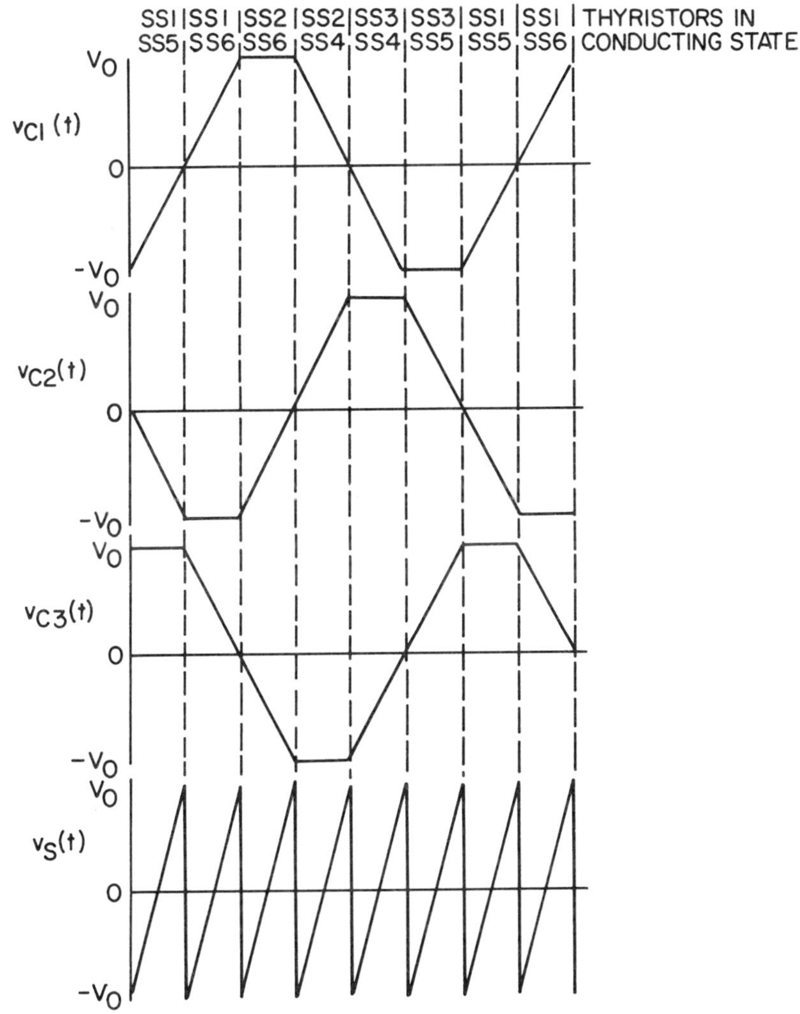

FIG. 4.6. Voltage waveforms on a three-phase inductor-converter bridge (ICB) when the load-coil current is zero.

mine $\Delta t'$. Kustom has investigated the limits of commutation through a complete energy transfer from the storage to the load at various different time (phase) differences between the converters. The investigation was done by checking the commutation conditions, interval by interval. The results for the three-phase ICB are shown in Figure 4.8. Analytical derivation of similar commutation threshold curves for the

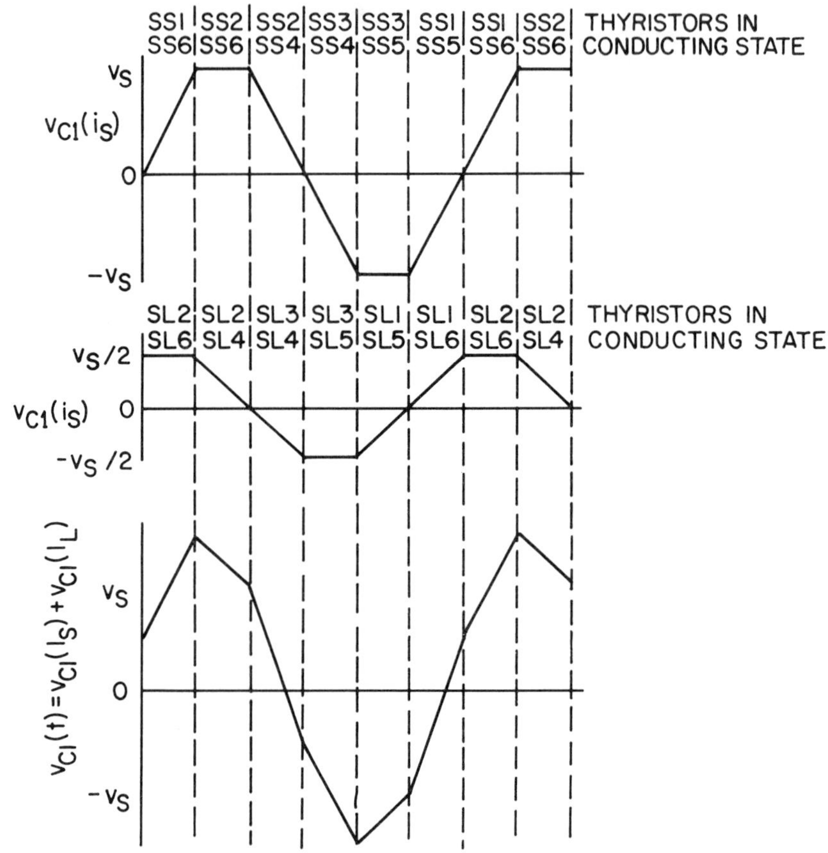

FIG. 4.7. Generation of capacitor waveform by superposition of storage-coil current and the load-coil current where load-coil current is half the storage-coil current and load switching leads storage by 60°, in the three-phase ICB.

one-phase ICB is demonstrated in chapter 5. In this figure, the limit of phase difference, for successful commutation, is plotted vs. the ratio i_L/i_S. For theoretically successful commutation, the storage/load phase difference must be between the two limits. It will be shown later that phase difference of $\phi = \pm 90°$ allows the maximum rate of energy transfer in the three-phase ICB. Therefore, for this ICB, no significant limitation is imposed by the commutation conditions in steady state. However, during transient switching, in which Δt is temporarily altered, commutation failures may still occur.

It is noteworthy that at the beginning of the energy transfer,

$i_L = 0$, the load inductor presents a high impedance to the load converter. Thus, the SCR holding current may not be established by the time the trigger pulse is terminated, and only small discontinuous pulses of current will flow in the load. Therefore, the energy transfer will not begin. The switched resistive branches in parallel with the inductors in Figure 4.9 provide the initial path for the holding current

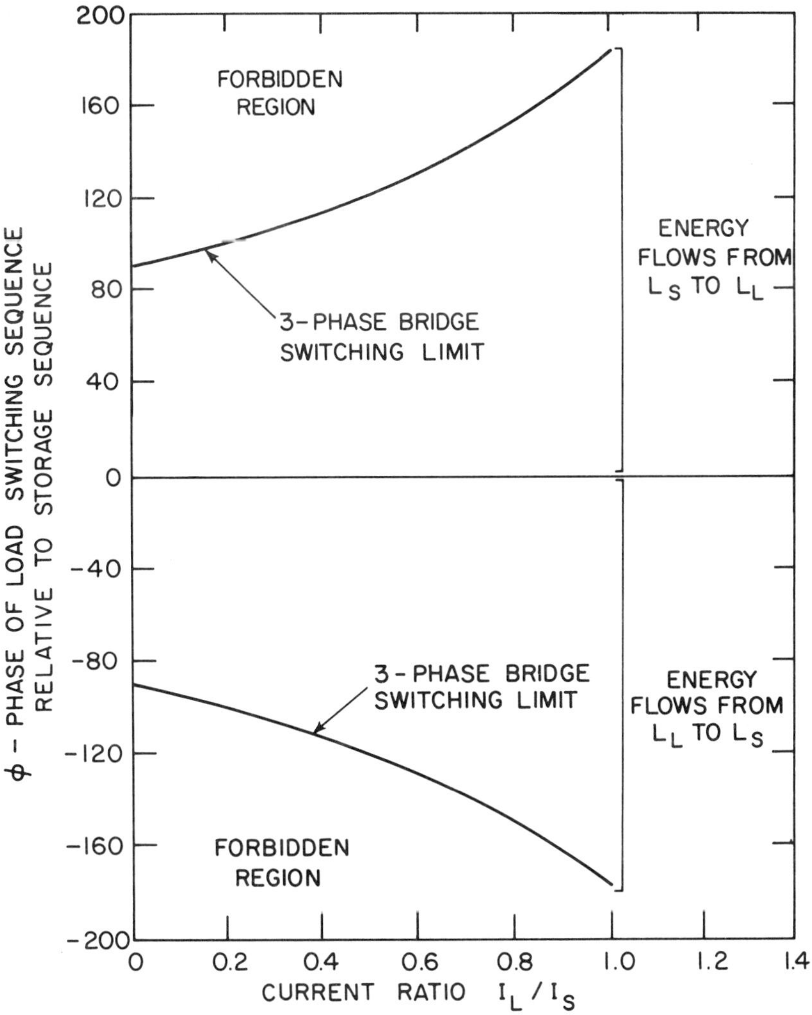

FIG. 4.8. Acceptable switching phases (ϕ) as a function of load-coil current divided by the storage-coil current in the three-phase ICB.

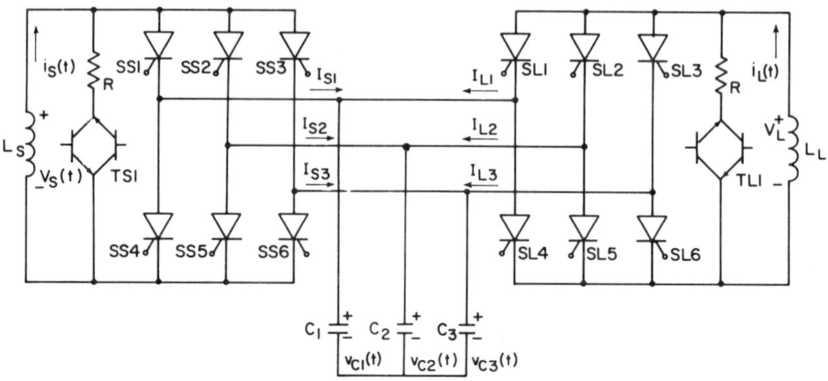

FIG. 4.9. Circuit diagram for a three-phase inductor-converter bridge (ICB) with startup shunt resistors.

of the SCRs. When the coil current safely exceeds the combined holding currents of the parallel SCRs of each phase, the lossy resistive paths are switched out by the series transistor switches, and efficient energy transfer commences. Alternative methods of solving this problem are presented in chapter 5.

4.2.2 Control of the ICB

The general concepts of ICB control are introduced in this section. These concepts are more rigorously treated in the following chapter. Net energy transfer from one ICB coil to the other takes place as a result of a phase difference between the two converters. For energy transfer to the load coil, the load-side converter is triggered with a phase lead relative to the storage converter. Figure 4.10 shows the capacitor and the coil voltage waveforms for the instance of $i_S = 2i_L$ and the load converter leading by one switching interval (60°). The instantaneous coil voltages are derived from the algebraic summing of the appropriate capacitor voltages, as dictated by each conduction mode. The figure shows that the storage and the load coils experience, respectively, a positive and a negative average voltage. By the polarity convention of Figure 4.5, this implies a net energy transfer to the load. A storage converter lead will result in a net energy transfer to the storage coil, as shown in Figure 4.11.

It is easy to see that the capacitor and inductor voltages depend on the level of coil currents, i_S and i_L; the relative phase (or timing, $\Delta t'$, in equation [4.16]) of the converters; and the frequency of con-

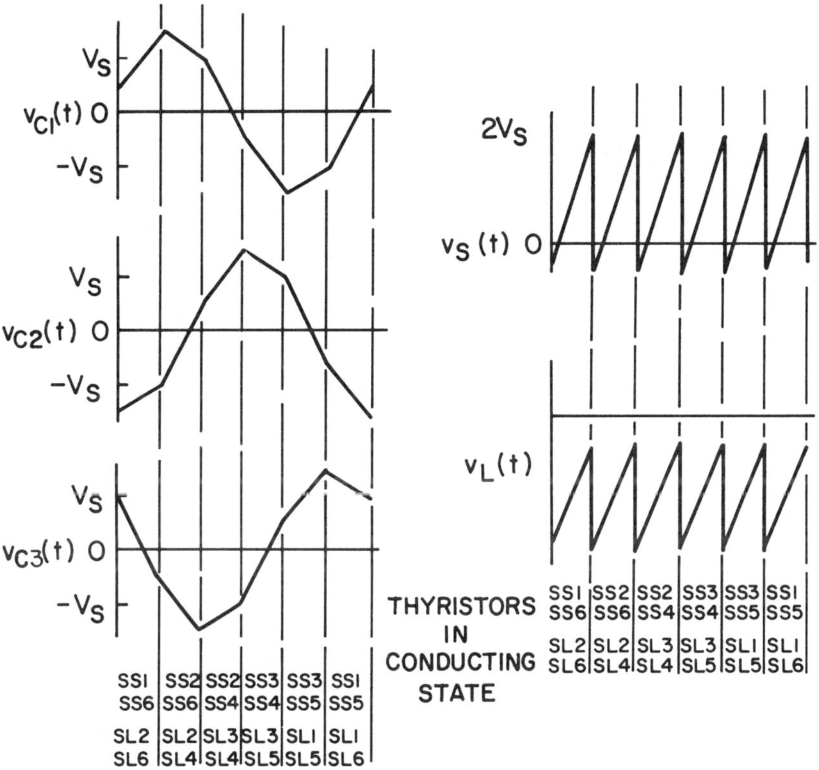

FIG. 4.10. Voltage waveforms on a three-phase inductor-converter bridge (ICB) when the storage-coil current is twice the load-coil current and the load-side switching sequence leads the storage side by 60°.

verter operation, which is proportional to Δt. Therefore, for real-time control, any one of these variables may be modulated.

4.2.2.1 Control by Frequency Modulation. Control of power by modulating the frequency of ICB operation was the original method suggested by Peterson and Mohan. In their method, the system is always operating at $+90°$ or $-90°$, depending on the desired direction of power. The level of power, according to these authors, is controlled in a feedback loop by modulating the frequency.

Change of frequency is effected by altering the switching intervals, Δt, from an initial value, Δt_1, to a final value, Δt_2. This change of conduction duration will alter the voltages of capacitors. Changes

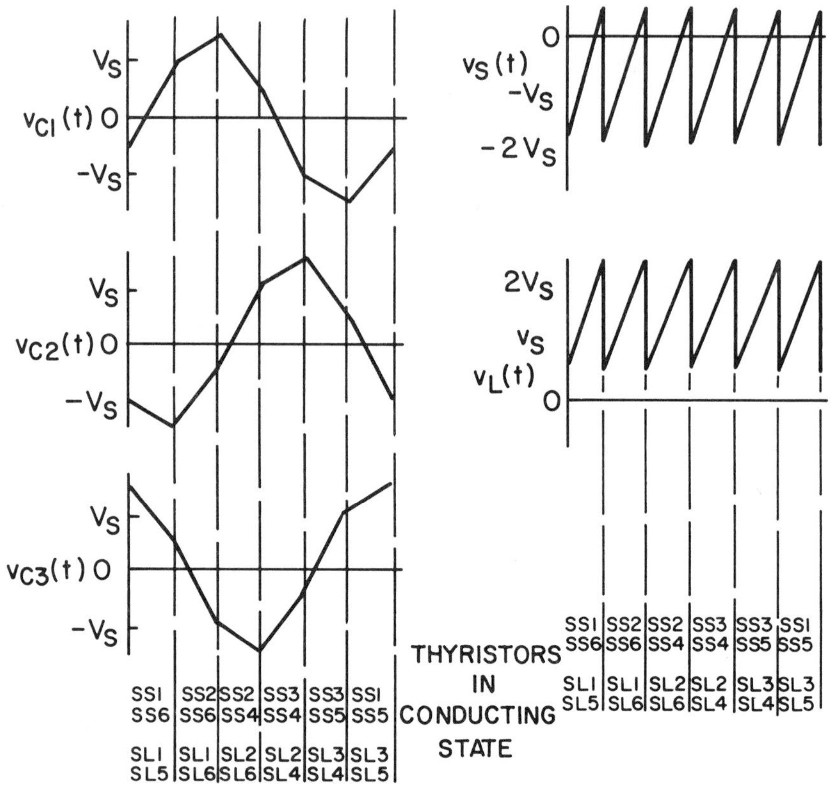

FIG. 4.11. Voltage waveforms on a three-phase inductor-converter bridge when the storage-coil current is twice the load-coil current.

in operation from one value of Δt to another must be done in a way that will not violate commutation conditions or upset the balance of the three-phase system. These disturbances may result in a failure in the ICB.

Kustom found a way of implementing frequency shift at any operating phase without disturbing the voltage balance. In chapter 5 a more fundamental theory of phase-frequency shifting, which was developed by Ehsani and Kustom, is presented that will encompass the above menthod as a special case.

4.2.2.2 Control by Phase Modulation. Control of power by modulating the phase was first suggested by Kustom. In order to alter the phase relationship between the two converters, one converter should temporarily operate at a higher or lower frequency than the other. This means that the switching interval, Δt, is temporarily shortened or lengthened for one or a few intervals of one converter and then returned to its normal duration. However, from equations (4.9) and (4.10) and the capacitor voltage waveforms of Figure 4.7, it is obvious that the capacitor voltage amplitudes will be altered by the change in Δt. Therefore, careful modulation of Δt is required so that the capacitor voltages will end up balanced and the commutation conditions will, at all times, be met.

In the suggested technique, any desired increment of change in phase is implemented in two half increments (split shifting), 180° apart from each other. Examples of changing the phase lead of the load converter from 60° to 90° are given in Figures 4.12, 4.13, and 4.14.

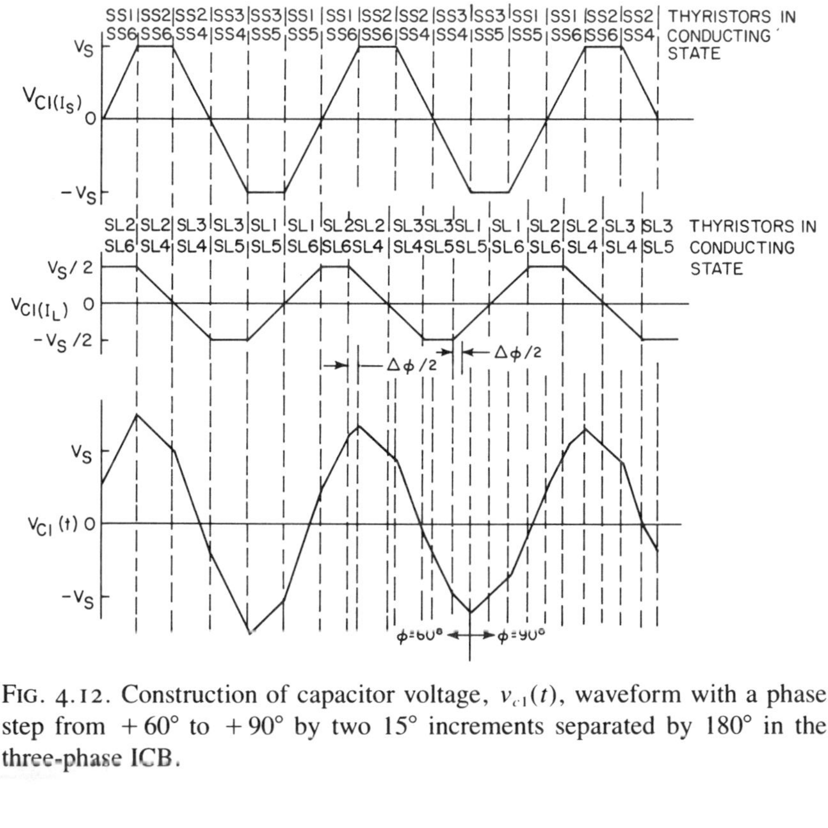

FIG. 4.12. Construction of capacitor voltage, $v_{c1}(t)$, waveform with a phase step from $+60°$ to $+90°$ by two 15° increments separated by 180° in the three-phase ICB.

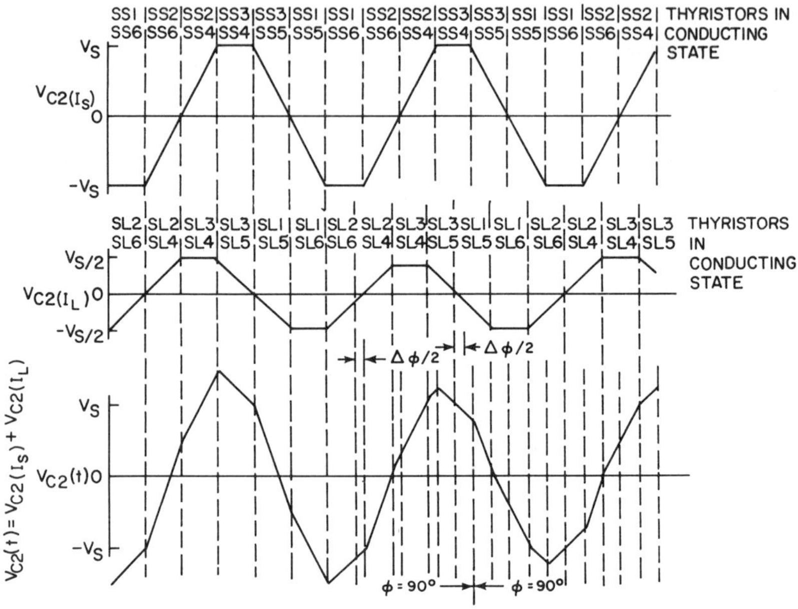

FIG. 4.13. Construction of capacitor voltage, $v_{c2}(t)$, waveform with a phase step from $+60°$ to $+90°$ by two 15° increments separated by 180° in the three-phase ICB.

The change is chosen to be implemented by shortening Δt on the load converter during conduction of SL2SL6 and, 180° later, SL3SL5.

Split shifting could have been used on the storage converter by lengthening Δt. The choice of the converter for split shift implementation depends on whether the temporary capacitor voltage alteration resulting from either the lengthened Δt or the shortened Δt may result in intolerable voltages. In a high-frequency system, shortened Δt may make Δt shorter than the SCR turn-off time and cause commutation failures.

A more fundamental treatment of phase shifting is presented in chapter 5.

4.2.3 Component Ratings

The method of evaluating component ratings for the ICB circuits is exemplified by the following evaluation of the three-phase ICB. The

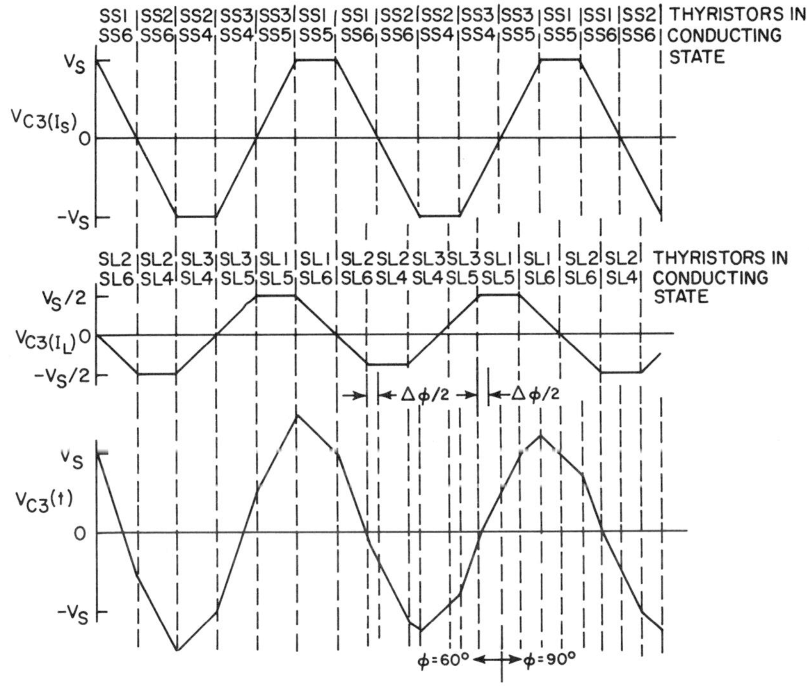

FIG. 4.14. Construction of capacitor voltage, $v_{c3}(t)$, waveform with a phase step from $+60°$ to $+90°$ by two $15°$ increments separated by $180°$ in the three-phase ICB.

installed value of capacitive energy storage, U_C, for the three-phase ICB is

$$U_C = \frac{3}{2} C V_m^2, \tag{4.17}$$

where V_m is the maximum capacitor voltage at any time in the cycle.

The highest capacitor voltage is generated at phase angles for which the peak values of the two superposition components are coincident in time. Therefore,

$$v_C = \frac{t_{sw}}{C} (I_S + I_L). \tag{4.18}$$

If I_L is assumed to be equal to kI_S, where k is a constant between 0 and 1, then

$$v_C = \frac{I_S t_{sw}}{C}(1 + k). \tag{4.19}$$

The currents I_S and I_L are related to each other through equation (3.32). If one assumes that U_0 is equal to $1/2(L_S I_0^2)$, equation (4.13) becomes

$$v_C = \frac{I_0 t_{sw}}{C}\left\{\frac{1 + k}{\sqrt{1 + k^2 L_L/L_S}}\right\}. \tag{4.20}$$

The maximum value of capacitor voltage occurs when k is equal to L_S/L_L. When L_S equals L_L, the value of installed capacitive energy is

$$U_C = \frac{3}{2}CV_m^2;$$

$$U_C = \frac{3}{2}C\left\{2\left(\frac{I_0^2 t_{sw}^2}{C^2}\right)\right\};$$

$$U_C = \frac{3I_0^2 t_{sw}^2}{C}. \tag{4.21}$$

The maximum forward and reverse holding voltages on each switch element of the bridge are twice the maximum capacitor voltage. This can be verified by considering the voltage across a given thyristor switch as the other two alternately enter the conducting state. The voltage across SS1, V_{SS1}, when SS2 is conducting is

$$v_{SS1} = -v_{C1} + v_{C2}, \tag{4.22}$$

and when SS3 is conducting,

$$v_{SS1} = -v_{C1} + v_{C3}. \tag{4.23}$$

The capacitor voltage waveforms when the operating phase is equal to zero degrees are the same as those shown in Figure 4.6 except that the peak voltage has the value of V_m in equation (4.20) with k equal to L_S/L_L. As can be seen in Figure 4.6, the value of V_{SS1} is $-2V_m$ from equation (4.22) at the instant that conduction switches from SS1SS6 and SL1SL6 to SS2SS6 and SL2SL6 and is $2V_m$ from equation (4.23) at the instant that conduction switches from SS3SS5 and SL3SL5 to SS1SS5 and SL1SL5.

The highest current a switch element must carry is equal to the highest current in the coil since the currents from the two coils are separately handled by the respective switching networks on either side of the capacitor. Furthermore, the switching element only conducts current for 1/3 of the total cycle; that is, the rms current rating per switch element is

$$I_{\text{rms}} = \sqrt{\frac{1}{T}\int_0^{T/3} I_0^2\,dt};$$

$$I_{\text{rms}} = I_0/\sqrt{3}. \tag{4.24}$$

4.2.4 Efficiency of Energy Transfer

Efficiency assessment of the ICB system is demonstrated by the following special case of the three-phase ICB. The current from the storage coil $i_S(t)$ flows through SS1, SS2, or SS3 and SS4, SS5, or SS6, contributing therefore to an energy loss proportional to the product of $2V_B$ and the number of series thyristors necessary to hold the voltage on the switch element. V_B is the barrier voltage of the solid-state switch junction. The load-coil current simultaneously flows through SL1, SL2, or SL3 and SL4, SL5, or SL6, also contributing to an energy loss proportional to the product of $2V_B$ and the number of series thyristors per switch element. The number of series elements per switch is the ratio $2V_m/V_T$, where V_m is equal to $I_0 t_{sw}\sqrt{(1+L_S L_L)}/C$ and V_T is the thyristor repetitive holding voltage rating.

The energy loss, U_e, is

$$U_e = \int_0^t \frac{4V_m V_B}{V_T} I_x \cos\theta(t)\,dt + \int_0^t \frac{4V_m V_B}{V_T} I_x \sqrt{\frac{L_S}{L_L}}\sin\theta(t)\,dt. \tag{4.25}$$

Equation (4.25) has the same form as equation (3.70). They are the same if $4V_m$ is replaced by $2V_0$. It should not be assumed that the three-phase ICB is less efficient simply because equation (4.25) has $4V_m$ in the product. The numerical value of $4V_m$ could be larger than, equal to, or smaller than the value of $2V_0$ for the SFC bridge, depending on the average coil voltage requirements throughout the cycle.

The above calculation gives an estimate of the ICB system losses associated with the switch conduction. For total converter loss estimation, the switching and other losses should be added to the conduction losses. These losses have been discussed in section 3.1.4.

4.3 The Multiphase ICB Circuit

The ICB concepts described thus far can be expanded to an m-phase system. A circuit diagram for an m-phase ICB is shown in Figure 4.15. As in the one-phase and three-phase ICB, the switching pulses are applied at equal time intervals, t_{sw}. Furthermore, the cyclic stepping of SS1 to SS2 . . . to SSm on the upper half of the bridge, and SS$m+1$ to SS$m+2$ . . . to SS$m+m$ and back to SS$m+1$ on the lower half of the bridge is followed. However, when $m > 3$, more than one switching pattern is possible. For example, the sequence could start with SS1SS$m+2$ to SS1SS$m+3$, etc., or SS1SS$m+3$ to SS1SS$m+4$ to SS2SS$m+4$. The various possible switching patterns can lead to symmetrical or asymmetrical voltage waveforms on the capacitors.

In the following sections the general characteristics of m-phase

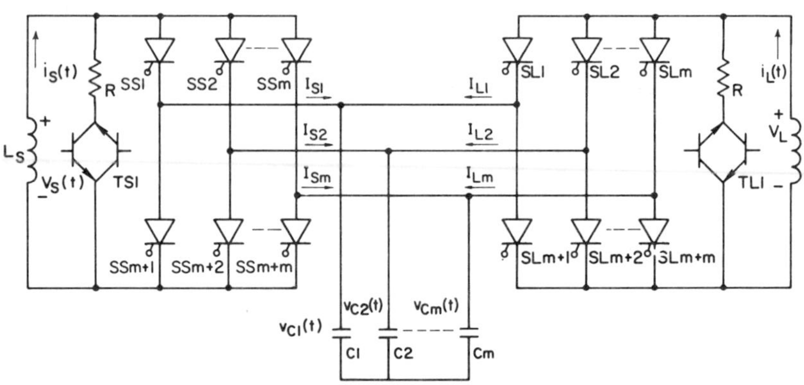

FIG. 4.15. Circuit diagram for the m-phase (two-m-pulse) inductor-converter bridge.

ICB systems are introduced. These circuits are then more exactly analyzed in the next chapter.

4.3.1 The *m*-phase ICB with Symmetrical Waveforms

In this section we will investigate the *m*-phase ICB circuits with an odd number of phases. The switching patterns which lead to symmetrical waveforms are the ones for which the sequence of conduction of switches is 1 and $(3m+1)/2$ to 1 and $[(3m+1)/2] + 1$ to 2 and $[(3m+1)/2] + 1$, and so on.

Typical capacitor voltage and phase current for an *m*-phase ICB with a switching pattern for symmetrical waveforms is shown in Figure 4.16.

The operating phase limits for successful commutations, corresponding to those shown in Figure 4.8 in a three-phase ICB, can also be found for *m*-phase ICB circuits. However, investigation has shown that the acceptable operating phase zone is broader than that for the three-phase ICB. As will be seen in the next chapter, phase angles between $+90°$ and $-90°$ provide all the control zone needed. Therefore, plot of phase limits for individual *m*-phase ICB systems is not very informative.

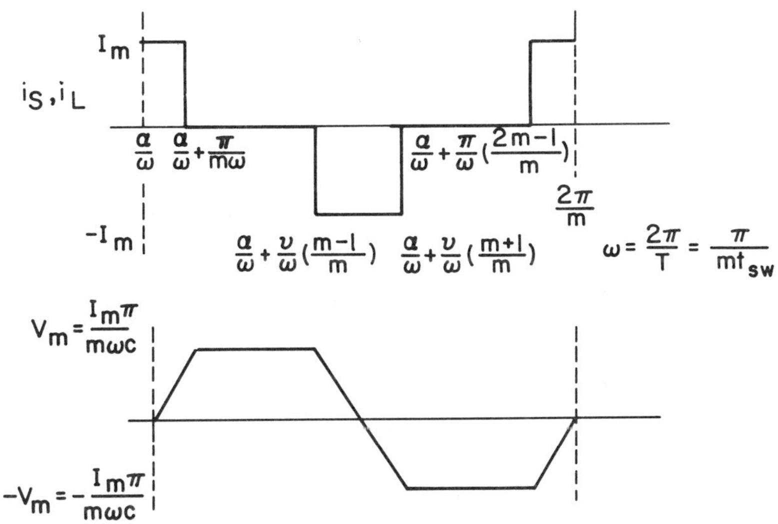

FIG. 4.16. Typical capacitor voltage and phase current waveforms for the *m*-phase inductor-converter bridge.

4.3.1.1 Active Control. Active control of the m-phase ICB is accomplished in the same manner as for the three-phase ICB. A leading phase angle of the load converter, relative to the storage converter, will cause a net power into the load coil and vice versa. Frequency control of power is also possible, but the sign of converter power is determined solely by the phase.

4.3.1.2 Component Ratings. The value of installed capacitive energy storage, U_C, for the m-phase ICB is

$$U_C = (m/2)CV_m^2, \tag{4.26}$$

where V_m is the maximum capacitor voltage. The maximum capacitor voltage in the m-phase bridge is the same value as for the three-phase bridge given in equation (4.20), where k is equal to L_S/L_L. The installed capacitive energy when L_S is equal to L_L is

$$U_C = \frac{mI_0^2 t_{sw}^2}{C}. \tag{4.27}$$

The rms current ratings on the individual switch elements decrease as $1/\sqrt{m}$. The total number of parallel thyristors needed for an m-phase bridge is proportional to the $\sqrt{m}$, again assuming that I_0 is much larger than the rms current ratings of available SCR devices so that large numbers of parallel combinations are needed for the m-phase bridge

$$I_{\text{rms}} = I_0/\sqrt{m}. \tag{4.28}$$

4.3.1.3 Efficiency. The energy loss during a transfer cycle for an m-phase bridge is described by equation (4.25).

4.3.2. The m-phase ICB with Asymmetrical Waveforms

The odd numbered m-phase ICB converters in which the order of switching sequence is different from those described in section 4.3.1 will lead to nonsymmetrical current and voltage waveforms. A typical nonsymmetrical current pattern on one of the m-phases is shown in Figure 4.17. The parameters r and q are positive integers. If $r = q$, the current waveform in Figure 4.17 will become symmetrical.

Asymmetrical modes are of little practical interest because for any given circuit investment they produce less maximum power capa-

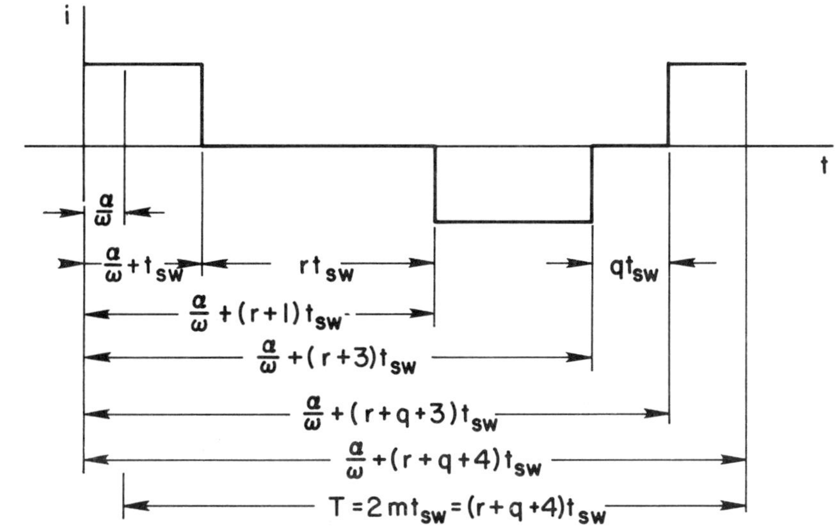

FIG. 4.17. Representative current waveforms for nonsymmetrical modes in the *m*-phase ICB.

bility. Furthermore, they produce higher harmonic content in the coil and capacitor voltage waveforms.

4.3.3 Even-Phased ICB Systems

Even-phased ICBs are the systems with converters consisting of 2, 4, 6, . . . phase currents. Symmetrical waveforms with successful commutations are possible in even-phased ICB circuits.

The two-phase ICB is simply a degenerate version of the one-phase ICB. In Figure 4.1 if the capacitor, C, is divided into two series capacitors, then the system is a two-phase ICB. The neutral point of the ac side voltages is the node between the capacitors.

However, nontrivial even-phased ICBs start with the four-phase ICB. The generalized form of the switching pattern used in the one-phase ICB may also be applied to the even-phased ICBs. If the required design conditions, described for the one-phase ICB, are met, commutation and symmetrical current waveforms will result. Figure 4.18 shows the storage side of a four-phase ICB and the resulting phase current.

The family of even phased ICBs completes the set of all *m*-phase ICB circuits.

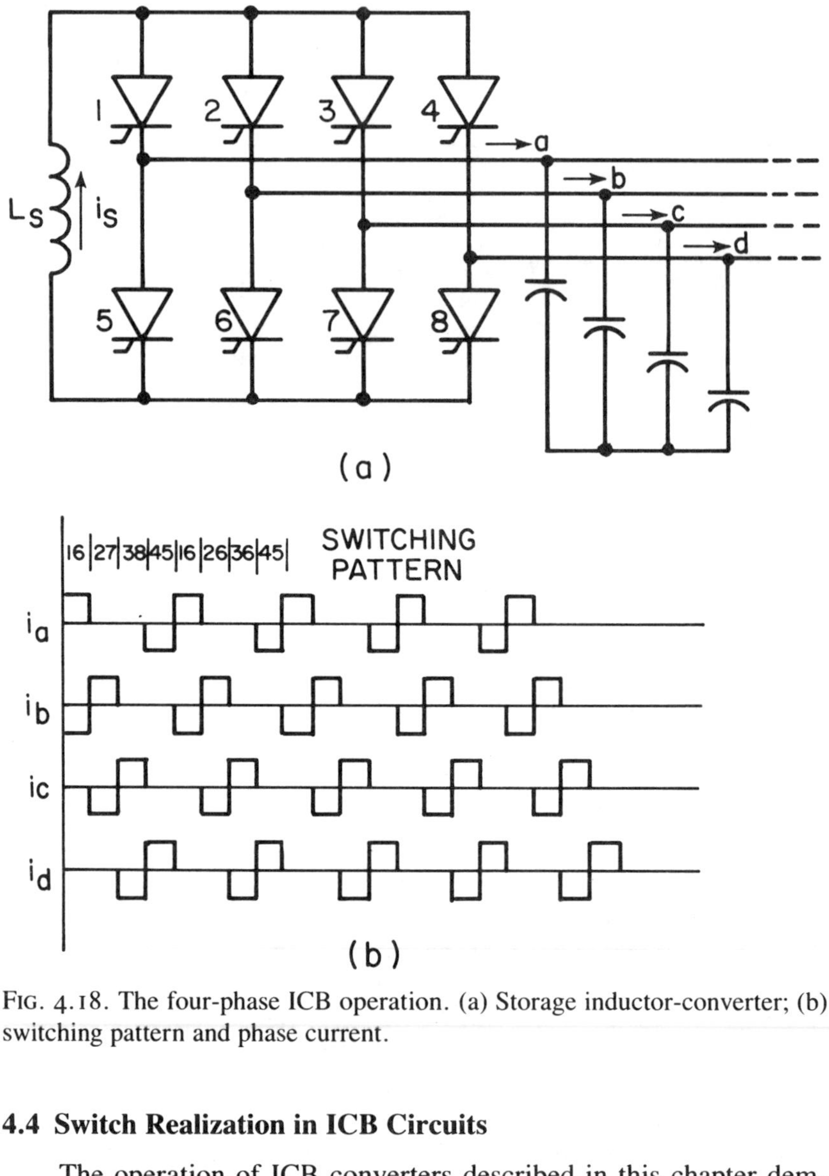

FIG. 4.18. The four-phase ICB operation. (a) Storage inductor-converter; (b) switching pattern and phase current.

4.4 Switch Realization in ICB Circuits

The operation of ICB converters described in this chapter demonstrates that if certain operating phase limits are observed, switch commutation voltage can always be available from the capacitors. These limits never impose any restrictions in the range of power control. The only additional requirement has been that the capacitors be appropriately precharged at the beginning of the energy transfer by an external means. However, it will be shown in the next chapter that capacitor precharging can also be accomplished by a special switching sequence on the storage converter at the beginning of the cycle.

With the availability of commutation voltage the SCR is the simplest switching device to use in the ICB circuit. The use of an SCR will also simplify the gate signal electronics, compared to that of other switching devices, and will provide the lowest losses for such high-current systems.

References

Ehsani, M., and R. L. Kustom. *Analysis of the Multiphase Inductor-Converter Bridge.* ANL/FPP Technical Memorandum No. 114. Argonne, Ill.: Argonne National Laboratory, August 1978.

Fuja, R. E.; R. L. Kustom; and R. P. Smith. "Results from a Model System of Superconducting Solenoids and Phase Shifting Bridge for Pulsed Power Studies for Proposed Tokamak EF Coils." 6th International Conference on Magnet Technology, Bratislava, Czechoslovakia, 1977.

Kustom, R. L. *Thyristor Networks for the Transfer of Energy Between Superconducting Coils.* Madison, University of Wisconsin Press, 1980.

———, R. Fuja, R. P. Smith, J. DeOlivares, A. Kellman, and T. J. Bauer. *The Use of Multiphase Inductor-Converter Bridges as Actively Controlled Power Supplies for Tokamak Coils.* Report No. ANL/FFP/TM-78. Argonne, Ill.: Fusion Power Program, Argonne National Laboratory, 1977.

———, R. B. Wehrle, R. P. Smith, and R. E. Fuja. "Argonne National Laboratory Energy Storage and Transfer Experimental Program." In *Proceedings of the 7th Symposium on Engineering Problems of Fusion Research.* Knoxville, Tenn., 1977.

Mohan, N., and H. A. Peterson. "Superconductive Inductor Storage and Converters for Pulsed Power Loads." In *Proceedings of International Pulsed Power Conference.* Texas Tech University, Lubbock.

Peterson, H. A., and N. Mohan. U.S. Patent 4,079,305, March 14, 1978.

———, N. Mohan, W. C. Young, and R. W. Boom. "Superconductive Inductor-Converter Units for Power Loads." In *Energy Storage, Compression, and Switching.* Ed. W. H. Bostick, V. Nardi, and O. S. F. Zucker. New York: Plenum Press, 1976.

CHAPTER 5

Analysis of the Inductor-Converter Bridge Circuits

In this chapter the inductor-converter bridge (ICB) is more rigorously studied. The time-averaged dynamic behavior of these circuits is mathematically derived. Then the methods of varying the operating phase and frequency by switching transients which will not upset the safe operation of the system are developed. Finally, some closed-loop control strategies which were developed for the ICB systems are discussed.

5.1 ICB Circuit Analysis by Fourier Method

The time-averaged dynamic behavior of the ICB systems was mathematically derived by Ehsani and Kustom. The general analysis is based on an m-phase ICB circuit.

5.1.1 Instantaneous Power in an m-phase ICB

The m-phase ICB of Figure 4.15 may be idealized by replacing the SCRs by their intended functional equivalent: on-off switches (Figure 5.1). This idealization implies that the SCR commutation conditions are not violated. It also implies that the SCR currents commutate in a time much shorter than either the conduction or blocking intervals. By design, the coil currents of the ICB will be almost unchanged during one converter cycle.* Therefore, viewing each converter from the m-phase ac lines, one will see a balanced m-phase rectangular current source system, Figure 5.2.† Each phase has an amplitude equal to the

*The basic design premise of the ICB system is that at any given time a very small fraction of the inductive energy is stored in the capacitors. Therefore, in one capacitor switching cycle the inductor currents change by a very small amount.

†The balanced nature of the converter phase currents is dictated by the steady-state switching pattern of the converters. For example, in a three-phase Graetz bridge triggering sequence, the negative sequence at all harmonics is zero. Also, all multiples of the third harmonic, including the zeroth multiple, are absent. This means that no zero sequence of any harmonic can be present, resulting in a balanced ac current.

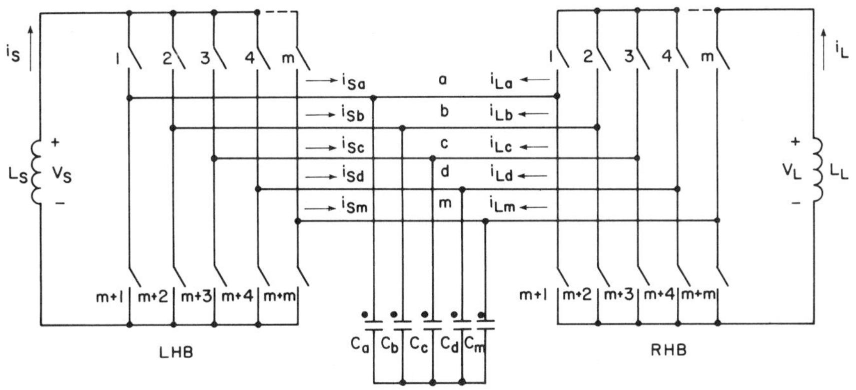

FIG. 5.1. Circuit diagram for an m-phase ICB.

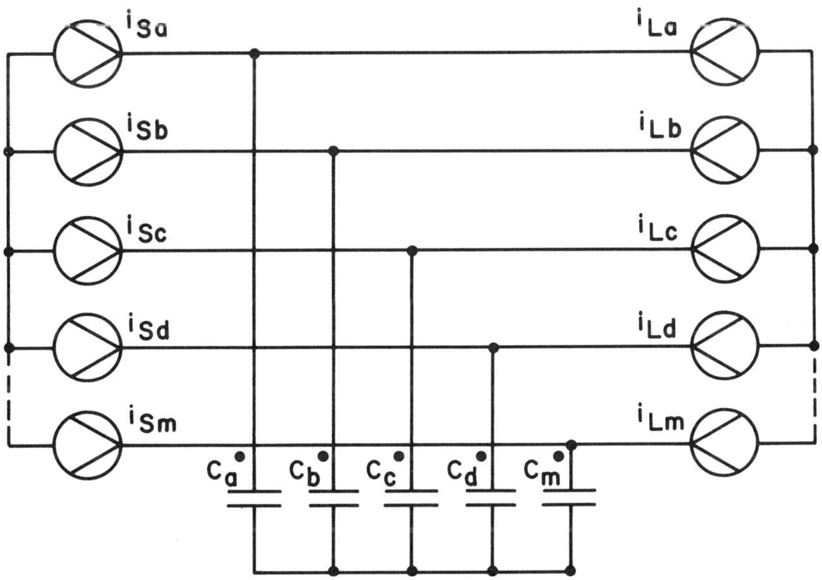

FIG. 5.2. Equivalent diagram for an m-phase ICB.

instantaneous coil current and is $2\pi/m$ radians displaced in time from the next phase.

The average power in one cycle of this equivalent circuit is the same as in the idealized ICB. Since the phases are identical, the net power per cycle delivered or received by either coil is divided equally among the phases. We may now proceed by calculating the power exchange between the two sides for one phase and multiplying the results by m for the total power.

Figure 5.3 shows the equivalent circuit of one phase (one line diagram). This equivalent circuit is the most basic unit of any m-phase ICB. It is interesting to note that this circuit is the dual of the one-phase equivalent circuit of a nonsalient pole synchronous generator connected to an ac system through its synchronous reactance. Here also the net energy flow is a result of phase difference between the two alternating sources. Calculations similar to those for the conventional ac system may be performed to derive the net power as a function of the phase difference. The only added complication is that the trigonometric calculations are done on the Fourier components of the square waves.

Representing the storage- and load-side currents in phase a, respectively, as

$$i_{S_a}(t) = \sum_{n=1}^{\infty} A_{nSa}\cos n\omega t + B_{nSa}\sin n\omega t \tag{5.1}$$

$$i_{La}(t) = \sum_{n=1}^{\infty} A_{nLa}\cos n\omega t + B_{nLa}\sin n\omega t, \tag{5.2}$$

the capacitor voltage resulting from the two currents will be

$$v_{Ca}(t) = V_{Ca0} + \frac{1}{C_a}\int_0^t [i_{Sa}(t) + i_{La}(t)]\, dt, \tag{5.3}$$

where V_{a0} is the initial voltage on C_a. The instantaneous power delivered by the storage converter is

$$p_{Sa}(t) = i_{Sa}(t) \cdot v_{Ca}(t). \tag{5.4}$$

The net or average power delivered by the storage in one cycle can now be calculated:

$$\langle p_{Sa}(t)\rangle = \frac{1}{t}\int_0^T p_{Sa}(t)\, dt, \tag{5.5}$$

where T is one converter cycle period.

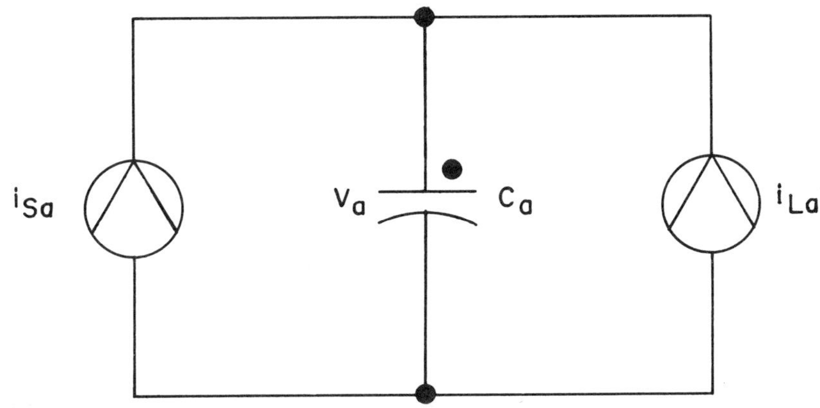

FIG. 5.3. One-phase equivalent of an m-phase ICB.

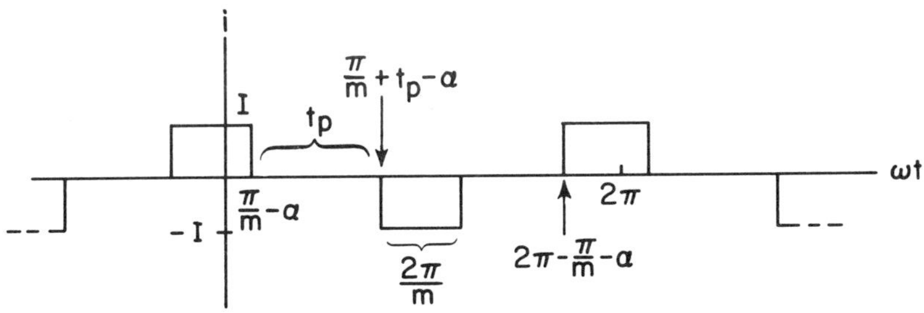

FIG. 5.4. General current waveform of each equivalent source in an m-phase ICB.

Substituting equations (5-1) through (5-4) in equation (5.5) and simplifying the trigonometric terms results in

$$\langle p_{Sa}(t)\rangle = \sum_{n=1}^{\infty} \frac{1}{2n\omega C_a}(B_{nSa}A_{nLa} - A_{nSa}B_{nLa}). \tag{5.6}$$

The remaining task is to calculate the Fourier coefficients in terms of the specific ICB currents under study. A general phase current waveform is shown in Figure 5.4. For most m-phase converter switching patterns, the length of each current pulse is $2\pi/m$. This waveform is drawn at an arbitrary time phase, ϕ, with respect to the reference. The separation between the positive and the negative current pulses is ϕ_p. This parameter depends on the number of converter phases and the chosen pattern of the converter switching. The significance of ϕ_p and the current pulse spacing will become apparent in the discussion of power equations.

If the storage current is taken as the phase reference ($\phi = 0$) and the load current at ϕ radians leading, the Fourier current components can be found as

$$A_{nSa} = \frac{2i_S}{n\pi}\sin\frac{n\pi}{m}\left[1 - \cos n\left(\frac{2\pi}{m} + \phi_p\right)\right] \tag{5.7}$$

$$B_{nSa} = -\frac{2i_S}{n\pi}\sin\frac{n\pi}{m}\sin n\left(\frac{2\pi}{m} + \phi_p\right) \tag{5.8}$$

$$A_{nLa} = \frac{2i_L}{n\pi}\sin\frac{n\pi}{m}\left\{\left[1 - \cos n\left(\frac{2\pi}{m} + \phi_p\right)\right]\cos n\phi - \sin n\left(\frac{2\pi}{m} + \phi_p\right)\sin n\phi\right\} \tag{5.9}$$

$$B_{nLa} = \frac{2i_L}{n\pi}\sin\frac{n\pi}{m}\left\{-\sin n\left(\frac{2\pi}{m} + \phi_p\right)\cos n\phi - \left[1 - \cos n\left(\frac{2\pi}{m} + \phi_p\right)\right]\sin n\phi\right\} \tag{5.10}$$

where i_S and i_L are the average coil currents over one period. Substitution of these Fourier components in equation (5.6) and simplification of the trigonometric terms will yield

$$\langle p_{Sa}(t)\rangle = \sum_{n=1}^{\infty}\frac{4i_Si_L}{n^3\pi^2\omega C_a}\left[1 - \cos n\left(\frac{2\pi}{m} + \phi_p\right)\right] \times \sin^2\frac{n\pi}{m}\sin n\phi. \tag{5.11}$$

The total net power delivered by the storage inductor is m times the value given in equation (5.11),

$$\langle p_S(t)\rangle = \sum_{n=1}^{\infty} \frac{4 i_S i_L m}{n^3 \pi^2 \omega C_a} \left[1 - \cos n\left(\frac{2\pi}{m} + \phi_p\right)\right] \times \sin^2 \frac{n\pi}{m} \sin n\phi. \tag{5.12}$$

Equation (5.12) describes the instantaneous average power delivered by the storage coil.

Several important conclusions may be drawn from this equation. First, the ϕ_p for maximum net power can be found by equating the quantity in the brackets to its maximum magnitude of 2:

$$\left| 1 - \cos n\left(\frac{2\pi}{m} + \phi_p\right)\right| = 2$$

$$\cos n\left(\frac{2\pi}{m} + \phi_p\right) = -1 \tag{5.13}$$

$$\frac{2\pi}{m} + \phi_p = \pi$$

$$\phi_p = \frac{(m - 2)\pi}{m}. \tag{5.14}$$

This is the general representation of the symmetrical switching pattern in odd-phased ICBs. Therefore, the symmetrical switching pattern, which was introduced in chapter 4, leads to the maximum rate of energy transfer for a given ICB design. This allows the control system to drive the ICB harder, in response to a step reference signal. In addition, SCR switching losses are reduced because more energy can be transferred per converter cycle. Substituting equation (5.14) in equation (5.12) will result in the general power equation for the odd-phased ICBs with symmetrical switching patterns:

$$\langle p_S(t)\rangle = \sum_{n=1}^{\infty} \frac{4 i_S i_L m}{n^3 \pi^2 \omega C} \left[1 - (-1)^n \right] \sin^2 \frac{n\pi}{m} \sin n\phi. \tag{5.15}$$

Second, the contribution of harmonics in the power equation is attenuated by the $1/n^3$ term. For the symmetrical switching pattern, the bracketed term in equation (5.15) is zero for even values of n. Therefore, even harmonics do not contribute to the net power. The next possible harmonic is $n = 3$, which can, at most, deliver 3.7% net power, compared to the fundamental frequency. In fact, it can be shown that at least 95% of the net power is delivered at the fundamental frequency. For the three-phase ICB this figure is better than 99%, owing to the absence of multiples of $n = 3$. This quantifies the error of the three-phase ICB analysis on the basis of the fundamental component, which has been done by Mohan.

It should also be noted that asymmetrical switching patterns not only reduce the fundamental frequency power but also allow power delivery, in both directions, by the even harmonics. For example, up to 12.5% of the power of the fundamental is delivered by the second harmonic. Power delivery by the higher harmonics, with their associated magnetic coupling losses and EMI, is expected to be undesirable.

Third, advancing the phase angle of the load bridge, ϕ, with respect to that of the storage bridge is the means of controlling the direction and the rate of energy transfer in the system. For ϕ positive, load leading, the net power delivered by the storage coil, $\langle p_S \rangle$, is positive, and vice versa. The angle of maximum net power, in general, may be found by maximizing $\langle p_s \rangle$ with respect to ϕ. Figure 5.5 shows a plot of the net power as a function of ϕ with the number of phases, m, as the parameter. It is seen that the net power peaks at a lower angle as the number of phases is increased. Note that because over 99% of the net power, in the three-phase ICB, is delivered at the fundamental frequency, this characteristic curve is almost purely sinusoidal. This similarity to the behavior of a three-phase synchronous generator is not surprising, considering the duality between their respective one-line diagrams.

Fourth, the converter operating frequency, ω, is another parameter by which the rate of energy transfer, but not its direction, may be controlled. Note that the power is inversely proportional to this variable, whereas the power control is almost proportional to sin ϕ. If control by ϕ only is chosen, the value of the operating frequency should be based on other design considerations such as the commutation losses and voltage ratings.

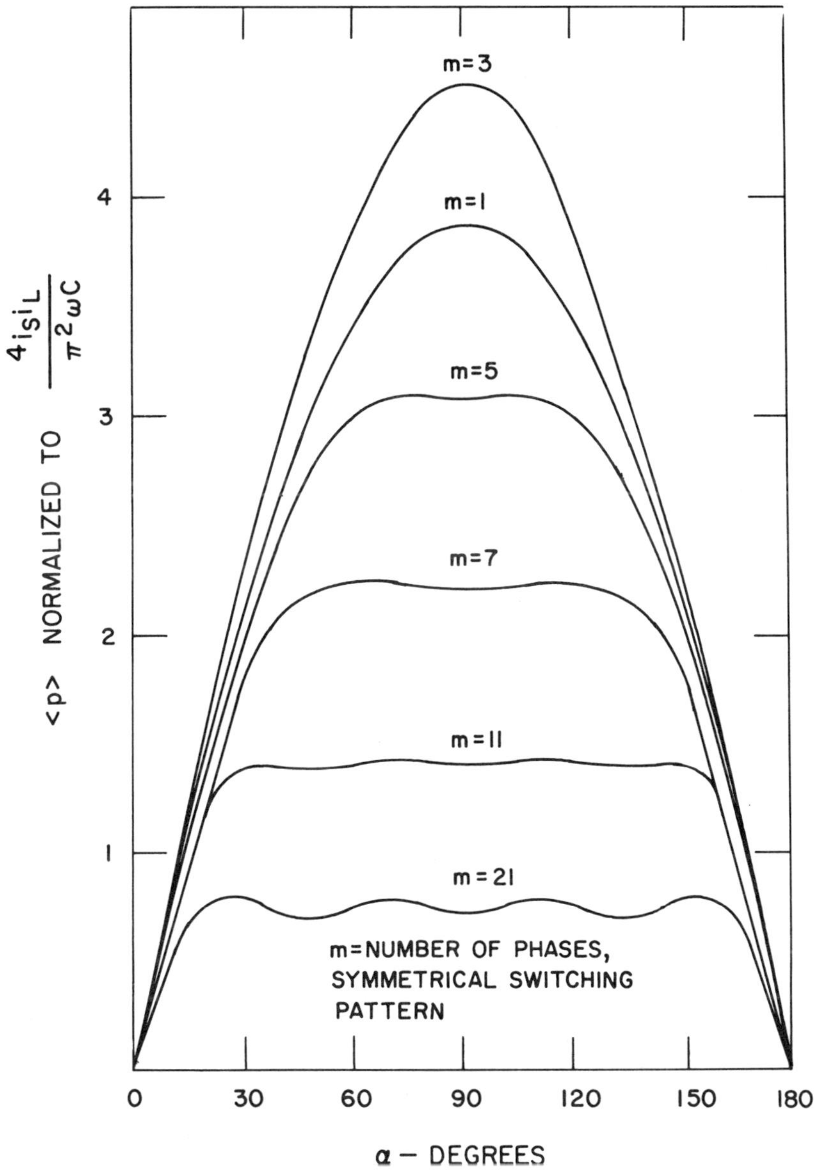

FIG. 5.5. Plot of $\langle p_s \rangle$ versus ϕ.

5.1.2 The Dynamic Quantities

The instantaneous average power was calculated with the assumption that the coil currents are static in one converter period.* Equation (5.12) may be rewritten as

$$\langle p_S \rangle(t) = k i_S(t) i_L(t), \tag{5.16}$$

where k represents the remaining terms of equation (5.12) and i_S and i_L are the average coil currents over one converter period.[†] The time variable, t, is meant to signify that the coil currents will change, slowly, over several converter periods. Conservation of inductive energy equations for the two coils may now be written as

$$\langle p_S \rangle(t) = -\frac{d}{dt}\left[\frac{1}{2} L_S i_S^2(t)\right] \tag{5.17}$$

$$\langle p_L \rangle(t) = \frac{d}{dt}\left[\frac{1}{2} L_L i_L^2(t)\right]. \tag{5.18}$$

The above simultaneous differential equations may be restated as

$$\dot{i}_S(t) + \frac{k}{L_S} i_L(t) = 0 \tag{5.19}$$

$$\dot{i}_L(t) - \frac{k}{L_L} i_S(t) = 0, \tag{5.20}$$

where $\dot{i}_S(t)$ and $\dot{i}_L(t)$ are time derivatives. Solving these equations with the initial conditions

$$i_S(0) = i_0 = \text{initial storage current}$$

$$\dot{i}_S(0) = \frac{k}{L_S} i_L(0) = 0 = \text{initial net voltage on the storage coil,}$$

*The approximation becomes exact as the frequency approaches infinity.

[†]Since the ratio of the capacitive to inductive stored energies is very small, the instantaneous and average coil currents are almost equal over one converter period.

will result in

$$i_S(t) = I_0 \cos \frac{k}{\sqrt{L_S L_L}} t \tag{5.21}$$

$$i_L(t) = I_0 \sqrt{\frac{L_S}{L_L}} \sin \frac{k}{\sqrt{L_S L_L}} t. \tag{5.22}$$

Equations (5.21) and (5.22) describe the instantaneous average coil currents in one complete lossless energy transfer from the storage to the load by their first quarter-cycle. Beyond this interval, the storage current equation indicates a reversal of current which is not possible because of the direction of conduction in the SCRs. Therefore, the energy transfer will terminate at the hypothetical moment of storage current zero crossing.

The dynamic net power equation may now be found by substituting equations (5.21) and (5.22) in equation (5.16). The result, after some trigonometric simplifications, is

$$\langle p_S \rangle(t) = \frac{1}{2} k I_0^2 \sqrt{\frac{L_S}{L_L}} \sin \frac{2k}{\sqrt{L_S L_L}} t. \tag{5.23}$$

The average coil voltages as functions of time may also be calculated directly from the currents. Using

$$\langle v \rangle(t) = L \frac{d}{dt} i(t), \tag{5.24}$$

we will find

$$\langle v_S \rangle(t) = -k I_0 \sqrt{\frac{L_S}{L_L}} \sin \frac{k}{\sqrt{L_S L_L}} \tag{5.25}$$

$$\langle v_L \rangle(t) = k I_0 \cos \frac{k}{\sqrt{L_S L_L}} t. \tag{5.26}$$

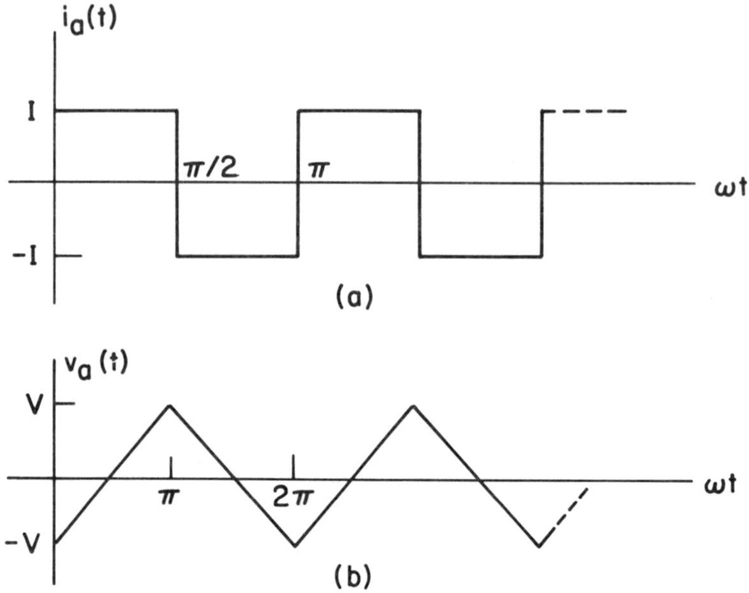

FIG. 5.6. Phase current and voltage of the one-phase ICB.

We can also show the functional dependence of the average coil voltages on the ratio of the two coil currents. This relationship for the load coil, $\langle v_L \rangle$, is found by eliminating the time variable between the equations (5.21), (5.22), and (5.26). The result is

$$\langle v_L \rangle = -kI_0/\sqrt{1 + L_L i_L^2 / L_S i_S^2}. \tag{5.27}$$

5.1.3 The Special Cases of One-phase and Three-phase ICBs

The one-phase ICB is the physical circuit realization of the one-phase equivalent circuit shown in Figure 5.3. The alternating current and voltage waveforms from each converter side of the one-phase ICB are shown in Figure 5.6. In this figure, the current amplitude is the instantaneous coil current of the corresponding side.

The Fourier components of the storage- and load-phase currents can be computed as

$$A_{nSa} = 0 \tag{5.28}$$

$$B_{nSa} = \frac{2i_s}{n\pi}[1 - (-1)^n] \tag{5.29}$$

$$A_{nLa} = \frac{2i_L}{n\pi}[1 + (-1)^{n+1}]\sin n\phi \tag{5.30}$$

$$B_{nLa} = \frac{2i_L}{n\pi}[1 - (-1)^n]\cos n\phi, \tag{5.31}$$

where ϕ is the advance angle of the load converter. Substituting equations (5.28) through (5.31) in equation (5.6) and simplifying results in

$$\langle p_S(t)\rangle = \sum_{n=1}^{\infty} \frac{4i_S i_L}{n^3\pi^2\omega C}[1 - (-1)^n]\sin n\phi. \tag{5.32}$$

Equation (5.32) could have been derived from equation (5.12) by substituting $m = 2$, $\phi_p = 0$, and $C = 2C$. The capacitor relationship exists because the phase capacitor in equation (5.32) is equivalent to one-half the phase to neutral capacitor in equation (5.12).

The operating phase for maximum power can be determined by differentiation as follows:

$$\frac{d}{d\phi}\langle p_S\rangle = 0.$$

The solution(s) is (are) the root(s) of the equation

$$\sum_{n=1}^{\infty} \frac{1}{n^2}[1 - (-1)^n]\cos n\phi = 0. \tag{5.33}$$

Therefore,

$$\cos\phi = 0 \rightarrow \phi = (q + 1)\frac{\pi}{2}, \qquad q = 1, 2, 3, \ldots .$$

i.e., maximum power is at $\phi = 90°$. Note that it is not easily discernible whether any other ϕ for maximum power exists. The unique ϕ for maximum power is calculated from the closed-form solutions derived in the next section.

The three-phase instantaneous average power is derived from equation (5.12) by substituting $m = 3$ and $\phi_p = \pi/3$:

$$\langle p_S(t) \rangle = k i_S(t) i_L(t)$$

$$k = \sum_{n=1}^{\infty} \frac{12}{n^2 \pi^2 \omega C} [1 - (-1)^n] \sin^2 \frac{n\pi}{m} \sin n\phi. \tag{5.34}$$

This equation will be used in the evaluation of the three-phase prototype ICB system. The extremal ϕ should be calculated from an equation similar to equation (5.33). The exact extremal ϕ will have to be derived from the closed-form solutions.

The Fourier method of analysis provides equations for the time-average waveforms of the ICB system in terms of the harmonics of the phase currents. This method shows both the positive and negative contributions of various harmonics in the energy transfer and control. The method can also be used to explicitly determine the choice of switching pattern that maximizes the power in a given m-phase ICB. An example of this switching pattern optimization was shown for the case of the three-phase ICB.

The Fourier method can also be used to study the harmonic content of various multiphase ICBs and different operating frequencies for a specific application. Therefore, known resonant frequencies of the coils can be avoided.

The Fourier method can establish accuracy of the analysis if the fundamental component approximation is used.

Finally, the Fourier method can be used to calculate the dynamic variables from trapezoidal phase current waveforms. This form of Fourier analysis would be necessary if the errors introduced by the assumption of instantaneous SCR commutation are unacceptable.

5.2 Circuit Analysis by Means of Square Functions

The dynamic quantities derived in the previous section depend on the variable k, which is defined in equation (5.34) for the three-phase ICB. Calculating k from the harmonic functions is always an approximation, albeit to any desired accuracy. Calculations with the dynamic functions are complicated by the presence of the infinite series, as in the ϕ calculation for maximum power. Furthermore, real-time computation of the dynamic quantities will take several milliseconds on contemporary microcomputers because of the presence of the infinite trigonometric series in the k term. This can be undesirable or intolerable for a converter system whose cycle time is less than one millisecond.

To avoid the computational difficulties of the infinite Fourier series, a specialized method of analysis for calculating the dynamic functions of the ICB has been developed by Ehsani. This method, which defines the phase currents in terms of unit step functions, produces closed-form algebraic expressions for the variable k. The derivation is simplified by defining two orthogonal functions and some common identities among these functions.

5.2.1 Quadrometrics

The discontinuous phase currents of the ideal ICB circuit may be mathematically expressed in terms of the superposition of several unit step functions, $u(t)$. Similarly, the phase capacitor voltage waveforms, which are the integral of the phase currents, may be expressed in terms of superposition of several ramp functions, $tu(t)$. Two general orthogonal functions, based on the step and ramp functions, have been developed to be used to express the ICB phase current and voltage waveforms, respectively.* These functions, which are named quadrometric functions, are

$$Sq\frac{1}{T}(t + t_0) \triangleq (-1)^m\left[u\left(\frac{t}{T} - n\right)\right.$$
$$- 2u\left(\frac{t}{T} - n - \frac{1}{2} + \frac{t_0}{T} - \frac{m}{2}\right)$$
$$\left. + 2u\left(\frac{t}{T} - n - 1 + \frac{t_0}{T} - \frac{m}{2}\right) - u\left(\frac{t}{T} - n - 1\right)\right] \tag{5.35}$$

$$Tr\frac{1}{T}(t + t_0) = (-1)^{m+1}\left[\left(\frac{t}{T} - n - \frac{1}{4} + \frac{t_0}{T} - \frac{m}{2}\right)\right.$$
$$\times u\left(\frac{t}{T} - n\right)$$

*To develop an appreciation before proceeding with the mathematical relationships presented in this section, the reader may first proceed to section 5.2.2.

$$- 2\left(\frac{t}{T} - n - \frac{1}{2} + \frac{t_0}{T} - \frac{m}{2}\right)u\left(\frac{t}{T} - n - \frac{1}{2} + \frac{t_0}{T} - \frac{m}{2}\right)$$

$$+ 2\left(\frac{t}{T} - n - 1 + \frac{t_0}{T} - \frac{m}{2}\right)u\left(\frac{t}{T} - n - 1 - \frac{t_0}{T} - \frac{m}{2}\right)$$

$$- \left(\frac{t}{T} - n - 1\right)u\left(\frac{t}{T} - n - 1\right)\Bigg]. \quad (5.36)$$

Quantities n, $m = 0, 1, 2, 3 \ldots$ where T is one period, t_0 is an arbitrary time shift, and $u(\cdot)$ is the unit step function. The integers n and m are unrelated to the n and m used previously and are found from the following relationships:

$$\frac{t}{T} - 1 < n < \frac{t}{T}$$

$$\frac{2t_0}{T} - 1 < m < \frac{2t_0}{T}. \quad (5.37)$$

These integers are merely for "bookkeeping" in general computations with the Sq and Tr functions at arbitrary instants of time and arbitrary time (phase) shifts. They do not appear in any specific calculation over one period as shown below.

Figure 5.7 shows the first period of Sq and Tr with time shift, $t_0 < T/2$. To describe this period, the general defining equations (5.35) and (5.36) simplify as follows:

$$0 \leqslant t \leqslant T \rightarrow n = 0$$

(the only allowable integers in equation [5.37]).

$$0 < t_0 < \frac{T}{2} \rightarrow m = 0$$

$$Sq\frac{1}{T}(t + t_0) = u\left(\frac{t}{T}\right) - 2u\left(\frac{t}{T} - \frac{1}{2} + \frac{t_0}{T}\right)$$

$$+ 2u\left(\frac{t}{T} - 1 + \frac{t_0}{T}\right) - u\left(\frac{t}{T} - 1\right) \quad (5.38)$$

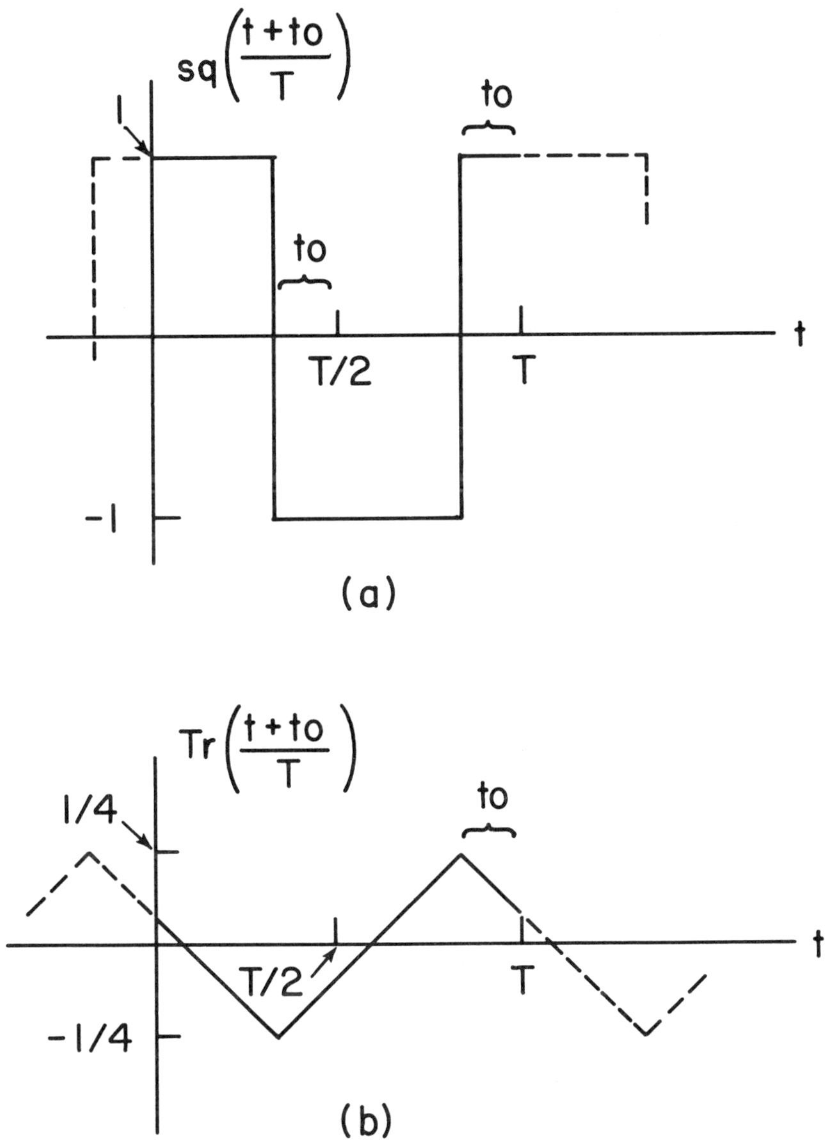

FIG. 5.7. Plot of $Sq(t)$ and $Tr(t)$ at an arbitrary phase shift.

$$\begin{aligned} Tr\frac{1}{T}(t + t_0) = -\Bigg[&\left(\frac{t}{T} - \frac{1}{4} + \frac{t_0}{T}\right)u\left(\frac{t}{T}\right) \\ &- 2\left(\frac{t}{T} - \frac{1}{2} + \frac{t_0}{T}\right)u\left(\frac{t}{T} - \frac{1}{2} + \frac{t_0}{T}\right) \\ &+ 2\left(\frac{t}{T} - 1 + \frac{t_0}{T}\right)u\left(\frac{t}{T} - 1 + \frac{t_0}{t}\right) \\ &- \left(\frac{t}{T} - 1\right)u\left(\frac{t}{T} - 1\right)\Bigg]. \end{aligned} \tag{5.39}$$

The first term in equation (5.38) is a unit step function starting at $t = 0$. The second term is a step function with a value of -2 and starting at $t = (T/2) - t_0$, etc.

Although the defining equations for *Sq* and *Tr* are somewhat cumbersome, they are only used to prove the basic identities for these functions. The calculations involving *Sq* and *Tr* are then performed using the identities. Some of these identities are listed below: indefinite integral relation,

$$\int Sq\frac{1}{T}(t + t_0)dt = -(T)\,Tr\frac{1}{T}(t + t_0); \tag{5.40}$$

definite integral relation,

$$\begin{aligned} \int_0^t Sq\frac{1}{T}(t + t_0)dt &= -(T)Tr\frac{1}{T}(t + t_0)\Bigg|_0^t \\ &= (T)\left[Tr\frac{1}{T}(t_0) - Tr\frac{1}{T}(t + t_0)\right] \\ &= (T)\Bigg[(-1)^{m+1}\left(-\frac{1}{4} + \frac{t_0}{T} - \frac{m}{2}\right) \\ &\quad - Tr\frac{1}{T}(t + t_0)\Bigg]; \qquad 0 \leq t \leq T \end{aligned} \tag{5.41}$$

multiples of $T/2$ shift,

$$Sq\frac{1}{T}(t) = (-1)^p Sq\frac{1}{T}\left(t + \frac{pT}{2}\right), \tag{5.42}$$

$$Tr\frac{1}{T}(t) = (-1)^p Tr\frac{1}{T}\left(t + \frac{pT}{2}\right), \tag{5.43}$$

$$p = 0, 1, 2, 3, \ldots;$$

symmetry about the time zero axis,

$$Sq\frac{1}{T}(t) = -Sq\frac{1}{T}(-t), \qquad \text{odd function,} \tag{5.44}$$

$$Tr\frac{1}{T}(t) = Tr\frac{1}{T}(-t), \qquad \text{even function;} \tag{5.45}$$

and orthogonality relation,

$$\int_{nT}^{(n+1)T} Sq\frac{1}{T}(t + t_1)\cdot Tr\frac{1}{T}(t - t_2)dt = \frac{-1}{T}\left[-2\left(t_1 - \frac{m_1T}{2}\right)^2 - T\left(t_1 - \frac{m_1T}{2}\right) + 4\left(t_1 - \frac{m_1T}{2}\right)\left(t_2 - \frac{m_2T}{2}\right) + T\left(t_2 - \frac{m_2T}{2}\right) - 2\left(t_2 - \frac{m_2T}{2}\right)^2\right],$$

$$\text{for } \frac{t_1}{T} - \frac{m_1}{2} \leqslant \frac{t_2}{T} - \frac{m_2}{T}$$

or

$$\int_{nT}^{(n+1)T} Sq\frac{1}{T}(t + t_1)\cdot Tr\frac{1}{T}(t - t_2)dt$$

$$= \frac{-1}{T}\left[2\left(t_1 - \frac{m_1T}{2}\right)^2 - T\left(t_1 - \frac{m_1T}{2}\right)\right.$$

$$-4\left(t_1 - \frac{m_1T}{2}\right)\left(t_2 - \frac{m_2T}{2}\right) + \left[T\left(t_2 - \frac{m_2T}{2}\right)\right.$$

$$\left.+2\left(t_2 - \frac{m_2T}{2}\right)^2\right], \qquad \text{for } \frac{t_2}{T} - \frac{m_2}{2} \leqslant \frac{t_1}{T} - \frac{m_1}{2} \tag{5.46}$$

$$m_1, m_2 = 0, 1, 2, 3, \ldots;$$

where m_1 and m_2 are evaluated from

$$\frac{2t_1}{T} - 1 < m_1 < \frac{2t_1}{T}$$

$$\frac{2t_2}{T} - 1 < m_2 < \frac{2t_2}{T}.$$

Orthogonality of *Sq* and *Tr* functions may be verified from equation (5.46) by setting $t_1 = t_2 = 0$, which will result in

$$\int_{nT}^{(n+1)T} Sq\frac{t}{T} \cdot Tr\frac{t}{T} dt = 0. \tag{5.47}$$

The cogent resemblance of the *Sq* and *Tr* functions to the sine and cosine functions, respectively, is evident. Calculations with *Sq* and *Tr* functions will be performed by invoking their trigonometric relationships.

5.2.2 The One-phase ICB

To illustrate the use of quadrometrics, the instantaneous average power of a two-phase ICB will be calculated. The two-phase ICB, Figure 4.1, is representable by the equivalent circuit of Figure 5.3. The source currents of this equivalent circuit, which have the form of Figure 5.6(a), may be described in terms of the *Sq* function as follows:

$$i_{Sa}(t) = i_S Sq\frac{t}{T} \tag{5.48}$$

$$i_{La}(t) = i_L Sq\frac{1}{T}(t + t_0), \tag{5.49}$$

where the storage converter is taken as the phase (or time) reference and the load converter is at a time lead to t_0 seconds. If the coil currents are assumed static over one period, the capacitor voltage of Figure 5.3 during one period is

$$\begin{aligned} v_C(t) &= \frac{1}{C}\int_0^t [i_{Sa}(\tau) + i_{La}(\tau)]d\tau + V_0 \\ &= \frac{i_S}{C}\int_0^t Sq\frac{\tau}{T}d\tau + \frac{i_L}{C}\int_0^t Sq\frac{1}{T}(\tau + t_0)d\tau + V_0, \end{aligned}$$

where V_0 is the initial capacitor voltage. The definite integral relation of equation (5.41) is used to evaluate the integrals

$$\begin{aligned} v_C(t) &= \frac{i_S}{C}\left[-(T)Tr\frac{\tau}{T}\right]_0^t + \frac{i_L}{C}\left[-(T)Tr\frac{1}{T}(\tau + t_0)\right]_0^t + V_0 \\ &= \frac{-i_S T}{C}\left[Tr\frac{t}{T} - Tr\frac{0}{T}\right] \\ &\qquad - \frac{i_L T}{C}\left[Tr\frac{1}{T}(t + t_0) - Tr\frac{0}{T}\right] + V_0. \end{aligned}$$

The value of Tr at zero is 1/4 which may be evaluated from equation (5.39) by setting $t = t_0 = 0$. Therefore,

$$\begin{aligned} v_C(t) &= -\frac{i_S T}{C}Tr\frac{t}{T} - \frac{i_L T}{C}Tr\frac{1}{T}(t + t_0) \\ &\qquad + \frac{i_S T}{4C} + \frac{i_L T}{4C} + V_0. \end{aligned} \tag{5.50}$$

The power delivered by the storage is

$$\begin{aligned} p_S(t) &= i_{Sa}(t)\cdot v_C(t) \\ &= -\frac{i_S^2 T}{C}Sq\frac{t}{T}Tr\frac{t}{T} - \frac{i_S i_L T}{C}Sq\frac{t}{T}Tr\frac{1}{T}(t + t_0) \\ &\qquad + i_S\left(\frac{i_S T}{4C} + \frac{i_L T}{4C} + V_0\right)Sq\frac{t}{T}. \end{aligned} \tag{5.51}$$

The average power delivered over any period is

$$\langle p_S(t)\rangle = \frac{1}{T}\int_{nT}^{(n+1)T} p_S(t)dt. \tag{5.52}$$

Substituting for the integrand from equation (5.51) results in

$$\begin{aligned}\langle p_S(t)\rangle = &-\frac{i_S^2}{C}\int_{nT}^{(n+1)T} Sq\frac{t}{T}Tr\frac{t}{T}dt \\ &-\frac{i_S i_L}{C}\int_{nT}^{(n+1)T} Sq\frac{t}{T}Tr\frac{1}{T}(t+t_0)dt \\ &+\left(\frac{i_S^2}{4C}+\frac{i_S i_L}{4C}+i_s V_0\right)\int_{nT}^{(n+1)T} Sq\frac{t}{T}dt.\end{aligned} \tag{5.53}$$

The first integral is zero because of the orthogonality of Sq and Tr, equation (5.47). The third integral is also zero because it averages the periodic Sq over one period. The second integral may be evaluated by using the identity of equation (5.46). The parameters of equation (5.46) can be evaluated by choosing the time shift $t_0 < T/2$ and noting that $t_1 = 0$, $t_2 = t_0$, and $m_1 = m_2 = 0$. The result is simply

$$\langle p_S(t)\rangle = \frac{i_S i_L}{C}\left(t_0 - 2\frac{t_0^2}{T}\right). \tag{5.54}$$

Equation (5.54) may also be expressed in terms of angular phase difference between the storage and load converters:

$$\langle p_S(t)\rangle = \frac{i_S i_L}{\omega C}\left(\phi - \frac{\phi^2}{\pi}\right), \tag{5.55}$$

where the phase variable is

$$\phi = 2\pi\frac{t_0}{T} = \omega t_0 \text{ radians}. \tag{5.56}$$

Equation (5.55) is the closed form of equation (5.32).

Note that the operating phase $0 \leqslant \phi \leqslant \pi$ covers the entire range of interest (see Figure 5.5).* Therefore, restricting the quadrometric analysis to $t_0 \leqslant T/2$ does not introduce any loss of generality.

5.2.3 Comparison with the One-phase Fourier Results

To verify the equivalence of equations (5.32) and (5.55), the latter will be expanded in its Fourier sine series to obtain the former. The two equations are repeated here for convenience:

$$\langle p_S \rangle_F = \sum_{n=1}^{\infty} \frac{4 i_S i_L}{n^3 \pi^2 \omega C}[1 - (-1)^n] \sin n\phi$$

$$\langle p_S \rangle_{Sq} = \frac{i_S i_L}{\omega C}\left(\phi - \frac{\phi^2}{\pi}\right).$$

Comparison of the two equations shows that for the power to be identical, we must have

$$\sum_{n=1}^{\infty} \frac{4}{n^3 \pi^2}[1 - (-1)^n] \sin n\phi = \phi - \frac{\phi^2}{\pi}. \tag{5.57}$$

Expanding the right-hand side in its Fourier sine series in the interval $[0, \pi]$† gives

$$\phi - \frac{\phi^2}{\pi} = \sum_{n=1}^{\infty} b_n \sin n\phi. \tag{5.58}$$

The values of the coefficients of b_n are

$$b_n = \frac{2}{\pi} \int_0^{\pi} \left(\phi - \frac{\phi^2}{\pi}\right) \sin n\phi \, d\phi$$

*For power reversal, $-\pi < \phi < 0$, identical results with a negative sign may be obtained.

†Since the absolute value of the average power delivered by the storage inductor is periodic in ϕ with period π, only the equivalence between the two expressions in the interval $[0, \pi]$, is of interest. Also, because equation (3.55) is valid in the above interval only.

$$= \frac{2}{\pi}\left[\frac{1}{n^2}\sin n\phi - \frac{\phi}{n}\cos n\phi\right.$$

$$\left.- \frac{1}{\pi}\left(\frac{2\phi}{n^2}\sin n\phi - \frac{n^2\phi^2 - 2}{n^3}\cos n\phi\right)\right]_0^{\pi}$$

$$= \frac{2}{\pi}\left[\frac{1}{n^2}\sin n\pi - \frac{\pi}{n}\cos n\pi - \frac{2}{n^2}\sin n\pi\right.$$

$$\left.+ \frac{n^2\pi^2 - 2}{n^3\pi}\cos n\pi + \frac{2}{n^3\pi}\right]$$

$$= \frac{2}{\pi}\left[-\frac{\pi}{n}\cos n\pi + \frac{\pi}{n}\cos n\pi - \frac{2}{n^3\pi}\cos n\pi + \frac{2}{n^3\pi}\right]$$

$$= \frac{4}{n^3\pi^2}(1 - \cos n\pi)$$

$$= \frac{4}{n^3\pi^2}[1 - (-1)^n]. \tag{5.59}$$

Therefore,

$$\phi - \frac{\phi^2}{\pi} = \sum_{n=1}^{\infty} \frac{4}{n^3\pi^2}[1 - (-1)^n]\sin n\phi$$

and the proof is completed.*

Equation (5.55) can be used to find the value of ϕ for the largest instantaneous average power transfer. By differentiation,

$$\frac{d}{d\phi}\langle p_S\rangle = \frac{i_S}{\omega C}\left(1 - \frac{2\phi}{\pi}\right) = 0.$$

*This proof is adequate for engineering utilization of the results. More rigorous justification would be needed from the pure mathematical point of view but with the same ultimate conclusion.

Solving for the extremal ϕ

$$\phi_{\max} = \frac{\pi}{2}. \tag{5.60}$$

The ϕ for maximum power can now be uniquely established as 90°.

5.2.4 The Three-phase ICB

The instantaneous average power in the three-phase ICB is recalculated in this section to further illustrate the use of quadrometric analysis and to obtain closed-form results.

As has been mentioned, the phase currents from each converter, in the three-phase ICB, have the form of Figure 5.4. This current waveform can be synthesized from two *Sq* functions, phased appropriately, by superposition, as shown in Figure 5.8. The phase currents from the storage and load sides, respectively, can be expressed as

$$i_{Sa}(t) = \frac{i_S}{2}\left[Sq\frac{1}{T}(t - t_0) + Sq\frac{1}{T}\left(t - t_0 + \frac{T}{6}\right)\right] \tag{5.61}$$

$$i_{La}(t) = \frac{i_L}{2}\left[Sq\frac{t}{T} + Sq\frac{1}{T}\left(t + \frac{T}{6}\right)\right], \tag{5.62}$$

where the timing difference between the storage and the load converters is shown as a lag for the storage. The capacitor voltage of Figure 5.3, during one period, is

$$\begin{aligned} v_a(t) &= \frac{1}{C}\int_0^t (i_{Sa} + i_{La})d\tau + V_0 \\ &= v_{Sa}(t) + v_{La}(t) + V_0 \\ &= -\frac{Ti_S}{2C}\left[Tr\frac{1}{T}(\tau - t_0) + Tr\frac{1}{T}\left(\tau - t_0 + \frac{T}{6}\right)\right]_0^t \\ &\quad -\frac{Ti_L}{2C}\left[Tr\frac{\tau}{T} + Tr\frac{1}{T}\left(\tau + \frac{T}{6}\right)\right]_0^t + V_0, \end{aligned} \tag{5.63}$$

where V_0 is the initial capacitor voltage. Here the definite integral identity of equation (5.41) is used, assuming static coil currents over one

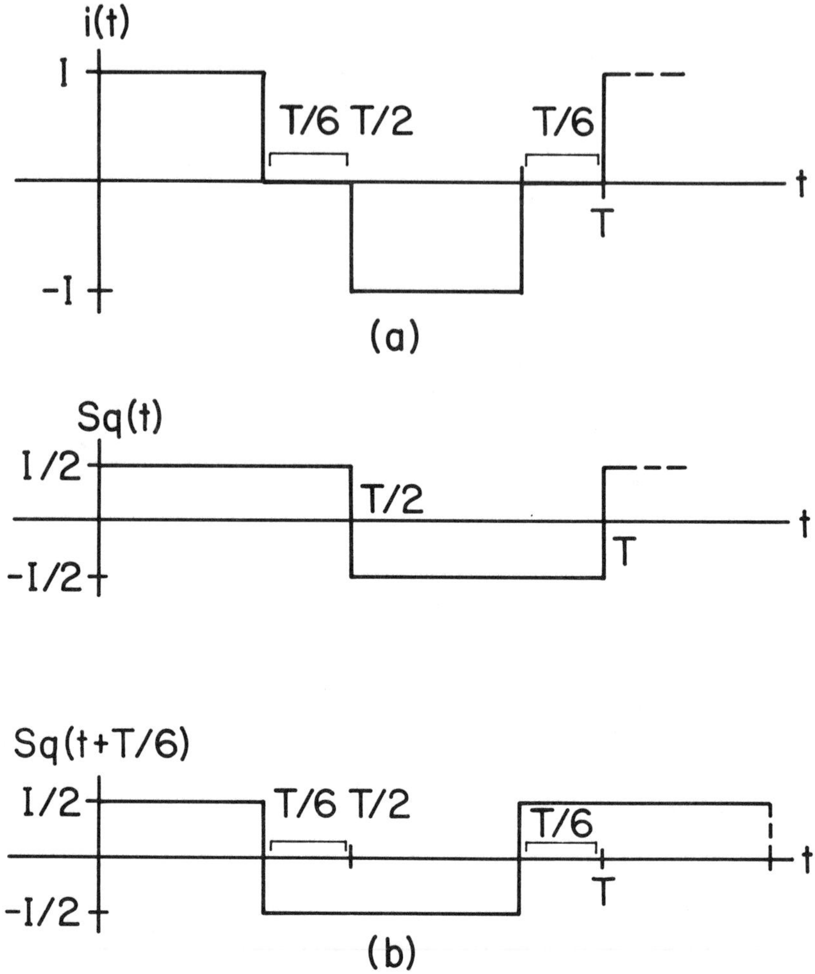

FIG. 5.8. (a) Phase current in a three-phase ICB; (b) decomposition of the phase current into two *Sq* functions.

period, T. The instantaneous power delivered by the storage side during one period is

$$\begin{aligned} p_{Sa}(t) &= i_{Sa}(t)v_a(t) \\ &= i_{Sa}(t)v_{Sa}(t) + i_{Sa}(t)v_{La}(t) + i_{Sa}(t)V_0. \end{aligned} \tag{5.64}$$

The average power over one period delivered by the storage side is

$$\begin{aligned}\langle p_{Sa}\rangle &= \frac{1}{T}\int_{nt}^{(n+1)T} i_{Sa}v_a\,dt \\ &= \frac{1}{T}\int_{nT}^{(n+1)T} i_{Sa}v_{Sa}\,dt + \frac{1}{T}\int_{nt}^{(n+1)T} i_{Sa}v_{La}\,dt \\ &\qquad + \frac{1}{T}\int_{nT}^{(n+1)T} i_{Sa}V_0\,dt. \end{aligned} \tag{5.65}$$

The third integral in equation (5.65) is the average of a periodic function over its period, which gives zero. The first integral can also be shown to give zero by orthogonality. However, it is physically apparent that this term results from periodic energy oscillation between the storage current source and the capacitor, and no net power can result. Thus, evaluating only the second term gives

$$\begin{aligned}\langle p_{Sa}\rangle &= \frac{1}{T}\int_{nT}^{(n+1)T} i_{Sa}(t)v_{La}(t)\,dt \\ &= -\frac{i_S i_L}{4C}\left\{\int_{nT}^{(n+1)T}\left[Sq\frac{1}{T}(t - t_0) + Sq\frac{1}{T}\left(t - t_0 + \frac{t}{6}\right)\right]\right. \\ &\qquad \times \left[Tr\frac{t}{T} + Tr\frac{1}{T}\left(t + \frac{T}{6}\right)\right]dt \\ &\quad - \int_{nT}^{(n+1)T}\left[Sq\frac{1}{T}(t - t_0) + Sq\frac{1}{T}\left(t - t_0 + \frac{T}{6}\right)\right] \\ &\qquad \left.\times\,[Tr(0) + Tr(0)]\,dt\right\}, \end{aligned} \tag{5.66}$$

where substitutions from equations (5.61) and (5.63) are made. The second integral in equation (5.66) has a periodic integrand and will result in zero. The non zero terms are

$$\langle p_{Sa} \rangle = -\frac{i_S i_L}{4C} \int_{nT}^{(n+1)T} \left[Sq\frac{1}{T}(t - t_0)\, Tr\frac{t}{T} \right.$$
$$+ Sq\frac{1}{T}(t - t_0)\, Tr\frac{1}{T}\left(t + \frac{T}{6}\right)$$
$$+ Sq\frac{1}{T}\left(t - t_0 + \frac{T}{6}\right) Tr\frac{t}{T}$$
$$\left. + Sq\frac{1}{T}\left(t - t_0 + \frac{T}{6}\right) Tr\frac{1}{T}\left(t + \frac{T}{6}\right) \right] dt. \quad (5.67)$$

From Figure 5.5, limiting the time shift of interest, $t_0 \leqslant T/2$, will not hinder generality. Now to avoid working with possible negative arguments, the identity of equation (5.42) is used to rewrite the Sq functions in terms of $T/2$ shift:

$$\langle p_{Sa} \rangle = \frac{i_S i_L}{4C} \int_{nT}^{(n+1)T} \left[Sq\frac{1}{T}\left(t - t_0 + \frac{T}{2}\right) Tr\frac{t}{T} \right.$$
$$+ Sq\frac{1}{T}\left(t - t_0 + \frac{T}{2}\right) Tr\frac{1}{T}\left(t + \frac{T}{6}\right)$$
$$+ Sq\frac{1}{T}\left(t - t_0 + \frac{2T}{3}\right) Tr\frac{t}{T}$$
$$+ Sq\frac{1}{T}\left(t - t_0 + \frac{2T}{3}\right) Tr\frac{1}{T}\left(t + \frac{T}{6}\right) dt. \quad (5.68)$$

Using the identity of equation (5.46) for each term of the integrand, the integration will break down to three intervals. Simplifying the results of each time interval, we will get the average power per phase delivered by the storage coil. The total average power is three times the single-phase power. The final results can be rewritten as

$$\langle p_s \rangle = \begin{cases} \dfrac{i_S i_L}{C}\left(2t_0 - 3\dfrac{t_0^2}{T}\right), & 0 \leqslant t_0 \leqslant T/6 \\[2ex] \dfrac{i_S i_L}{C}\left(3t_0 - \dfrac{6t_0^2}{T} - \dfrac{T}{12}\right), & T/6 \leqslant t_0 \leqslant T/3 \\[2ex] \dfrac{i_S i_L}{C}\left(t_0 3\dfrac{t_2 0}{T} + \dfrac{T}{4}\right), & T/3 \leqslant t_0 \leqslant T/2. \end{cases} \tag{5.69}$$

Equation (5.69) is the closed form of the three-phase average power if the substitution $t_0 = \phi/\omega$ is made. It is evident that the closed form of k in equation (5.69) can be written in three segments of 60° each. Figure 5.9 shows a plot of this average power as a function of $\phi_0 = t_0/T$. This is the same as the $m = 3$ characteristics in figure 5.5. The extremal ϕ can easily be calculated from the differentiation of the middle expression in equation (5.69). This ϕ is found to be uniquely equal to 90°.

The above analysis is based on the assumption that the SCR commutations take place instantaneously. In reality the counterpulse current in the conducting SCR will rise as an underdamped resonant current in the commutation loop. During the interval of commutation this current rise closely resembles a linear ramp function. Therefore, the rectangular-phase current waveforms shown in Figures 5.4 and 5.6 could be trapezoidal to account for the commutation.

The approximation involved in assuming instantaneous commutation may be put in perspective by the following discussion. Each half-cycle of a trapezoidal phase current would contain a rising interval, t_{rise}, and a conduction interval, $t_{\text{cond.}}$. The rise time is fundamentally limited by the di/dt of the on-coming SCR. The minimum conduction time is fundamentally limited by the turn-off time, t_q of the off-going SCR. In a typical fast rise time SCR, $di/dt = 800$ A/μs. This means that, for a typical conduction level of 1000 A per SCR, the rise time is about 1 μs. The fastest turn-off time in today's large SCRs is about 10 μs. Therefore, for this worst-case example, the ratio of conduction to rise time will be $t_{\text{cond.}}/t_{\text{rise}} = 10$. It is observed that in this case it may be marginally acceptable to approximate such a trapezoidal wave by a square wave. However, for efficient energy flow in

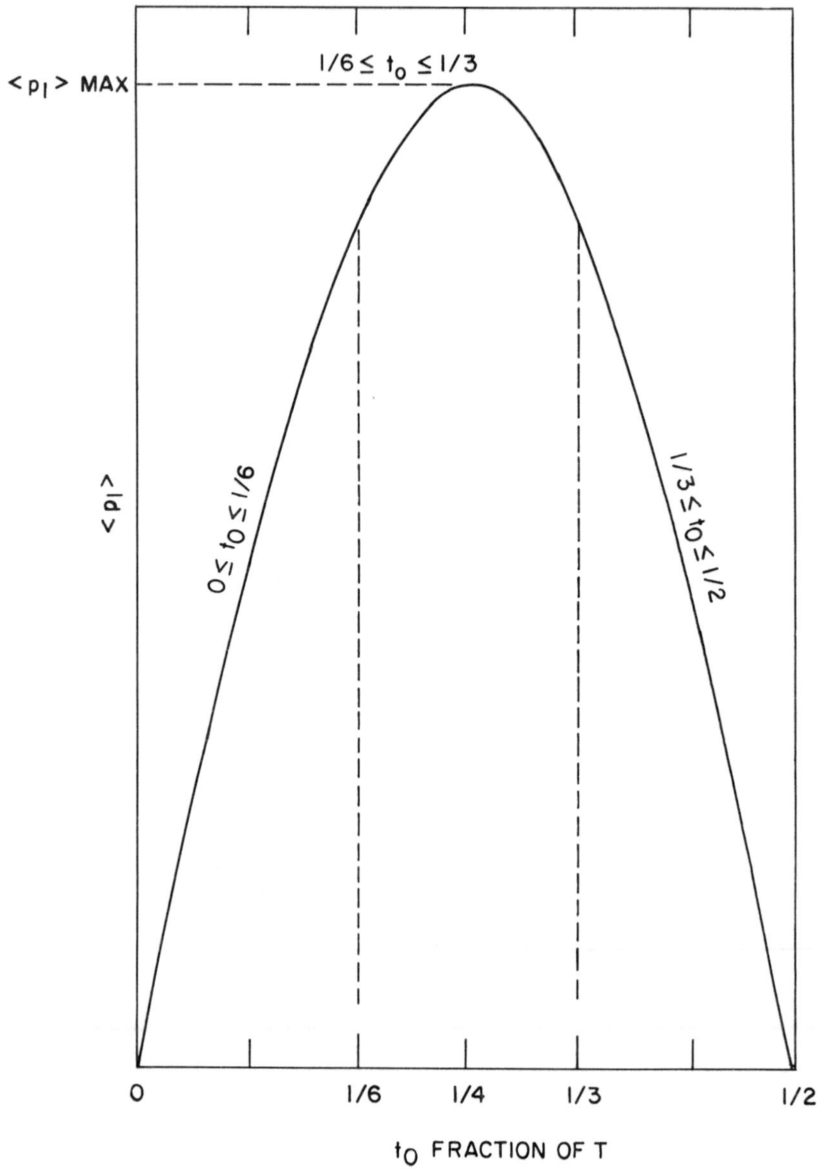

FIG. 5.9. The average power versus t_o in a three-phase ICB.

the ICB, $t_{\text{cond.}}$ is many times t_q. Thus, the square wave approximation becomes quite acceptable.

Reference

Ehsani, M., and R. L. Kustom. *Square Function Analysis of the Inductor-Converter Bridge*. Argonne National Laboratory Report ANL/FPP/TM-118. Argonne, Ill.: Argonne National Laboratory, March 1979.

CHAPTER 6

Methods of Phase and Frequency Shifting in the Inductor-Converter Bridge Circuits

The derivation of the dynamic current, voltage, and power equations in the previous chapter assumes flawless operation of the converters and successful commutation of the SCRs. Furthermore, the equations were derived for constant phase and frequency of operation. Control of the load voltage, current, or power requires active control of the phase and frequency. These in turn vary the parameter k in the dynamic equations (5.21), (5.22), (5.23), (5.25), and (5.26).

Under steady-state operation the coil current is commutated from one SCR to another at uniform time intervals. However, when rapidly changing the operating phase and/or frequency, the intervals between some of the gate pulses must be varied over some period of time between the old and new operating conditions. These changes are defined as transient switching sequences, and the period of time over which the transient switching sequence is accomplished is defined as the transient switching period or interval. Transient switching sequences can produce overvoltages or voltage imbalances on the capacitors if done incorrectly. In this chapter we present the fundamental concepts of acceptable alterations in ICB switching. These ideas are then embodied in a mathematical method which may be used systematically to derive the acceptable switching transients.

6.1 General Constraints on Phase and Frequency Shifting

If the number of commutation steps, N, in the transient switching period is relatively large, then the capacitor voltage imbalances generated will be very small. Under these conditions no special transient switching sequences need to be used. Consider, for instance, the effect of a slow increase in frequency taking n steps. If the starting interval is t_s and the interval corresponding to the new frequency is δt smaller than t_s, then each succeeding commutation step is decreased by the amount $\delta t/n$; i.e.,

$$
\begin{aligned}
t_1 &= t_s - \delta t/n, \\
t_2 &= t_s - 2\delta t/n, \\
&\ \cdot \\
&\ \cdot \\
&\ \cdot \\
t_n &= t_s - \delta t.
\end{aligned}
$$

The amount by which an individual capacitor voltage changes during each commutation interval will be successively smaller. This is illustrated in Figure 6.1. If the voltages are calculated step-by-step, it is apparent that each successive value of V^+ or V^- is $6I(\delta t)/nC$ smaller and the imbalance $(V^+ - V^-)/2$, is $2.5I(\delta t)/nC$. Thus, the capacitor voltage slowly reduces to the value corresponding to the balanced condition at the higher frequency, and the level of imbalance slowly decreases starting with an order of magnitude equal to $[2.5(\delta t)/nt_s] \times 100\%$.

Similar conditions result in slowly changing the relative phase on the ICB or the relative coil currents as described by equations (5.21) and (5.22). In an energy transfer cycle where the current is transferred from L_S to L_L, or vice versa, with fixed phase and frequency operation, the correct balanced capacitor voltages shown in Figures 4.3, 4.4, 4.6, and 4.7 are generated throughout the cycle by this adiabatic process.

For most active control systems the response time for changing the phase and frequency must be within a few cycles rather than the larger number of cycles needed for adiabatic balancing of the converter. Special transient sequences are developed for this purpose.

The choice of applying the transient on storage, load, or both converter sides, and the kind of transient, is made on the basis of other constraints, as will be discussed later. For frequency shifting, the transient switching is implemented on both converter sides, because the two converters must operate at the same frequency at all times. Transient switching sequences for phase shifting and frequency shifting can be combined into a composite transient switching sequence. This is often necessary when frequency shifting is implemented. Frequency

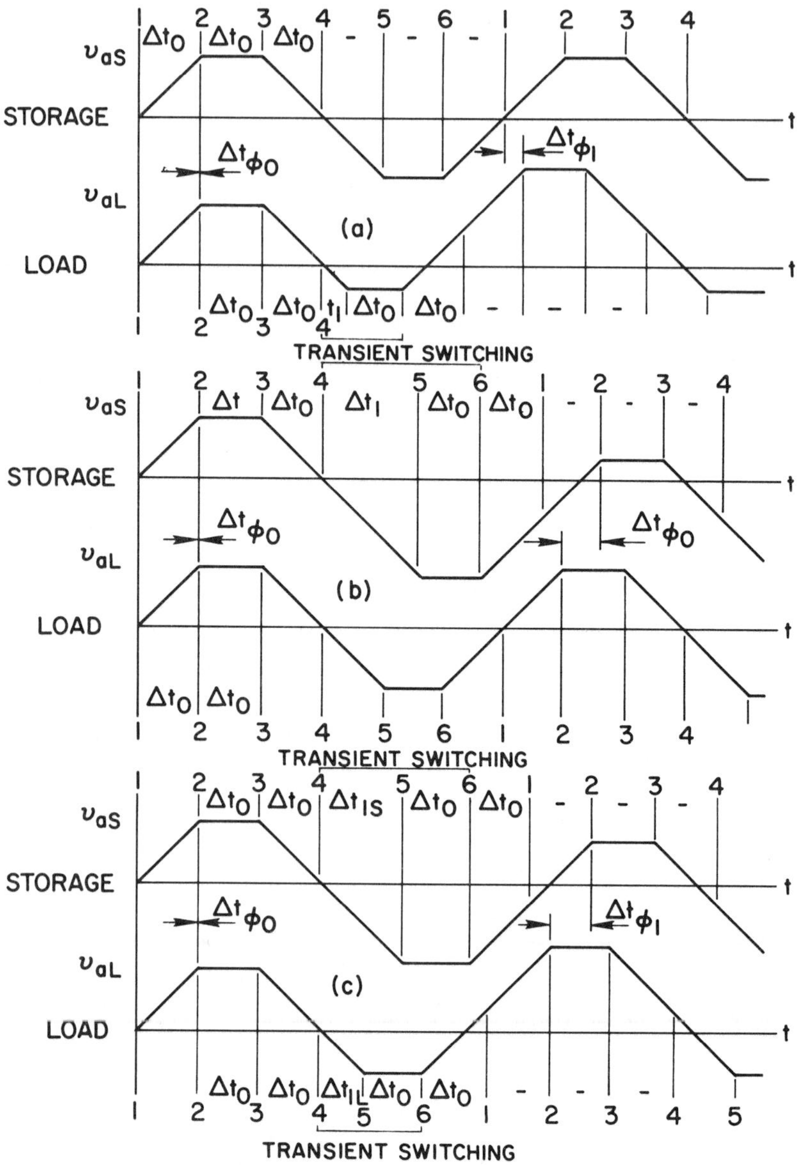

FIG. 6.1. Three ways of inducing a two-step transient switching for the same phase control. Events 1→6 represent the six switching actions in the conventional Gratz bridge. Note the development of mode voltage bias.

shifting will alter the phase relationship that existed at the original frequency.*

A permissible transient switching for any m-phase ICB should satisfy six constraints imposed by the ICB circuit and the controlling system.

1. The transient switching should produce the desired phase relationship.

2. The transient switching should produce the desired frequency of operation.

3. The transient switching should not cause any capacitor overvoltages. Such overvoltages may be introduced if, for example, too long an interval is imposed between the switching events. Overvoltages can cause capacitor and/or SCR breakdown.

4. The transient switching should not cause any SCR commutation failures in either converter. Commutation success is independent of the operating frequency within the minimum and maximum design frequencies of a given ICB system. Commutation failure threshold curves as a function of the steady-state phase relationship have been numerically computed for the three-phase ICB (Figure 4.8) and will be mathematically derived for the one-phase ICB in this chapter.

5. The transient switching should not cause any phase voltage imbalances after the termination of the transient. It should be noted that in steady-state switching, balance will always be maintained in an m-phase ICB, because such circuits cannot produce zero sequence or negative sequence currents.† However, careless transient switching such as that shown in Figures 6.1, 6.2, and 6.3 will produce phase voltage imbalances.

6. The transient switching should occur in as short a period of

*A frequency change will automatically alter the operating phase. This is because the relative timing delay between the converters at the new frequency will mean a new phase difference:

$$\phi_0 = (360°)\frac{\Delta t_\phi}{T_1},$$

$$\phi_1 = (360°)\frac{\Delta t_\phi}{T_2},$$

$$\phi_0 \neq \phi_1$$

†One-phase voltage balance is defined as $\int_t^{t+T} v(t)\,dt = 0$, where T is one converter period.

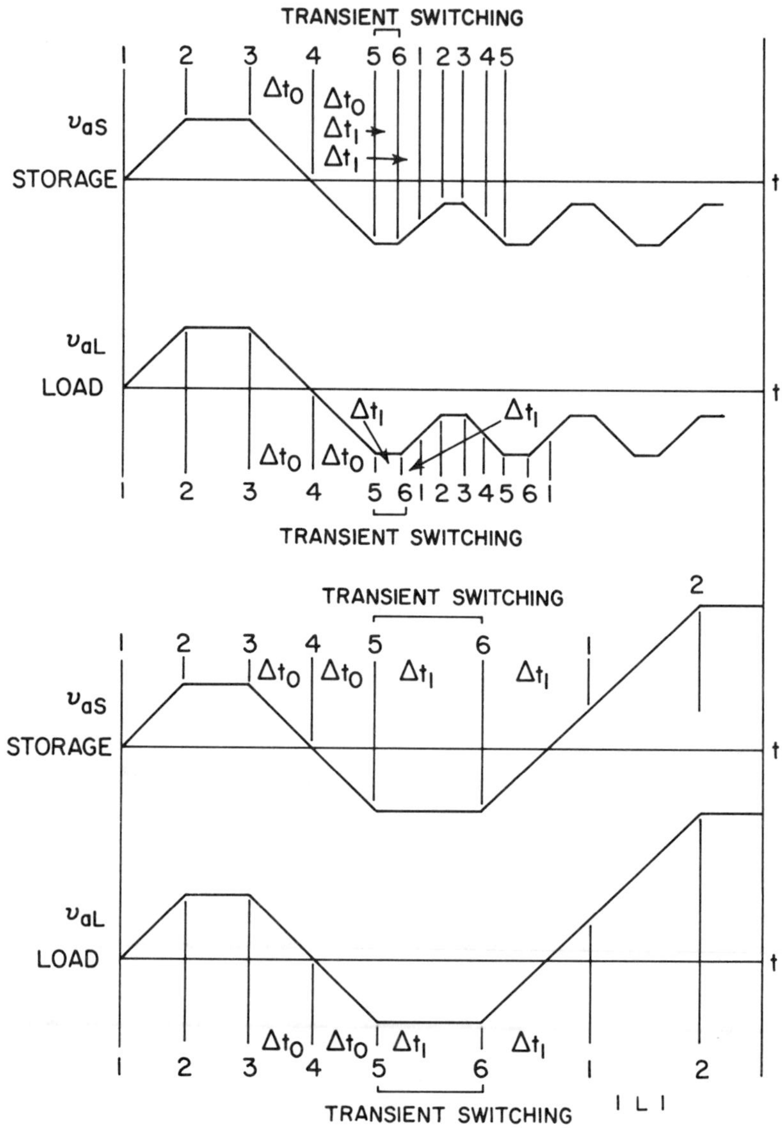

FIG. 6.2. One-step transient switching for frequency control: (a) frequence increase; (b) frequency decrease. Note the development of dc voltage bias.

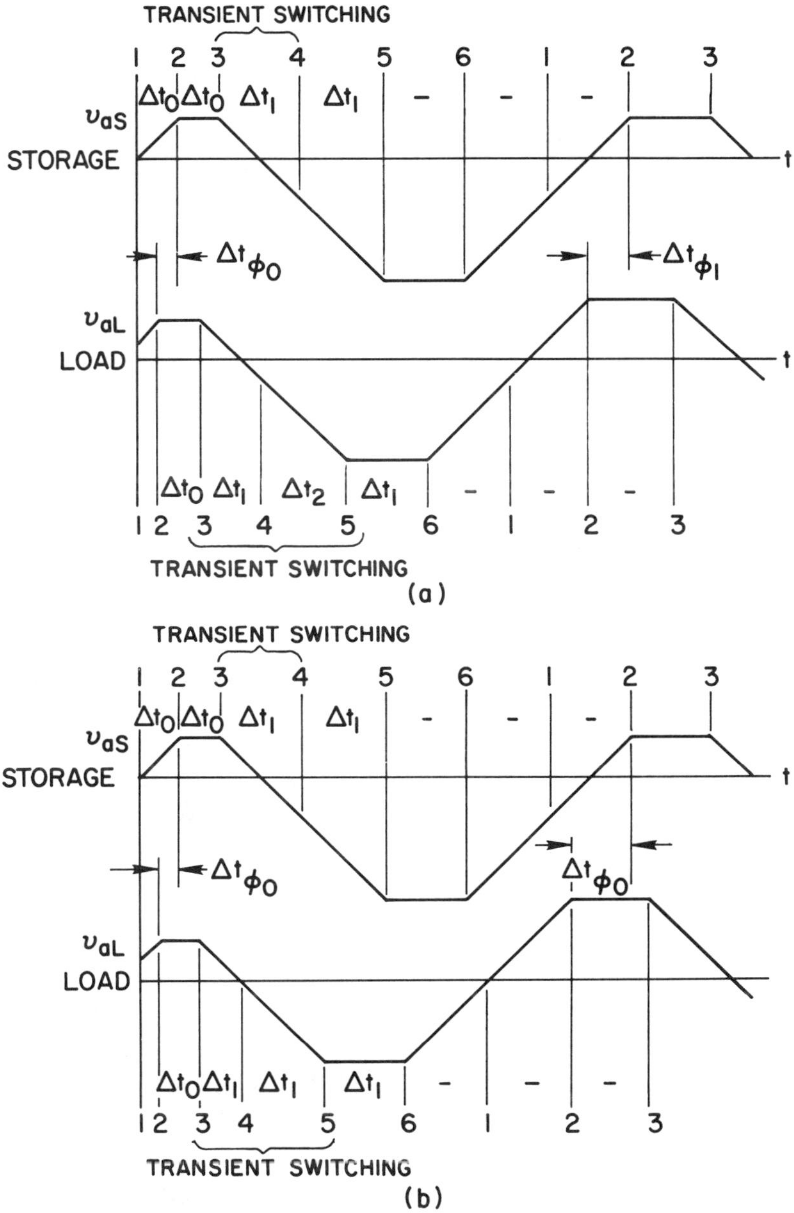

FIG. 6.3. Two-step transient switching for phase-frequency control: (a) maintaining constant phase while changing frequency; (b) changing both phase and frequency. Note the development of dc voltage bias.

time as possible. This will improve the system response time to the reference control signal.

6.2 Phase Shifting in the One-Phase ICB

The shortest transient switching sequence for phase shifting in the one-phase ICB, without producing capacitor voltage imbalance, is a three-step one. If a phase shift corresponding to a Δt_ϕ time shift between the storage and load converters is required, the three transient steps of the sequence are

$$\Delta t_1^{(3)} = \Delta t_1 - \Delta t_\phi/2 \tag{6.1}$$

$$\Delta t_2^{(3)} = \Delta t_1 - \Delta t_\phi/2 \tag{6.2}$$

$$\Delta t_3^{(3)} = \Delta t_1 \tag{6.3}$$

where $\Delta t_1^{(3)}$, $\Delta t_2^{(3)}$, and $\Delta t_3^{(3)}$ symbolize the first, second, and third steps of a three-step sequence and Δt_1 is the steady-state switching time interval. The subscripts 1, 2, and 3 indicate the first, second, and third steps of the sequence, and the superscript (3) indicates the number of commutation steps in the transient switching sequence. Phase shifting can also be affected by transient steps which are longer than the steady-state step. However, longer switching intervals produce temporary overvoltages on the capacitor. This may cause system overvoltage failures if the ICB is running close to its minimum safe frequency. Therefore, transient switching steps which are longer than the steady-state step will not be considered further. Figure 6.4 shows an example of 0° to 90° phase advancing of the storage converter. Note that the shortened transient steps should always be applied to the converter that is advancing in time. Therefore, decreasing shifts in ϕ (90° → − 90°), are implemented on the storage converter and vice versa. Examination of capacitor commutation voltage polarities in Figure 6.4 shows ample commutation voltage available during the transient. This suggests that a phase shift larger than 90° could be implemented at once. Figure 6.5 shows one such example where $i_L/i_S \simeq 0$ and a phase shift $\phi = 90°$ → − 90° is implemented in three transient steps. Note that all commutation events have the correct capacitor voltage, v_C, polarity, albeit a small one, during and after the shift. The conclusion is that if interval-shortening transients are used, phase shifts to and from the threshold of safe operating angles may be implemented in three steps. The

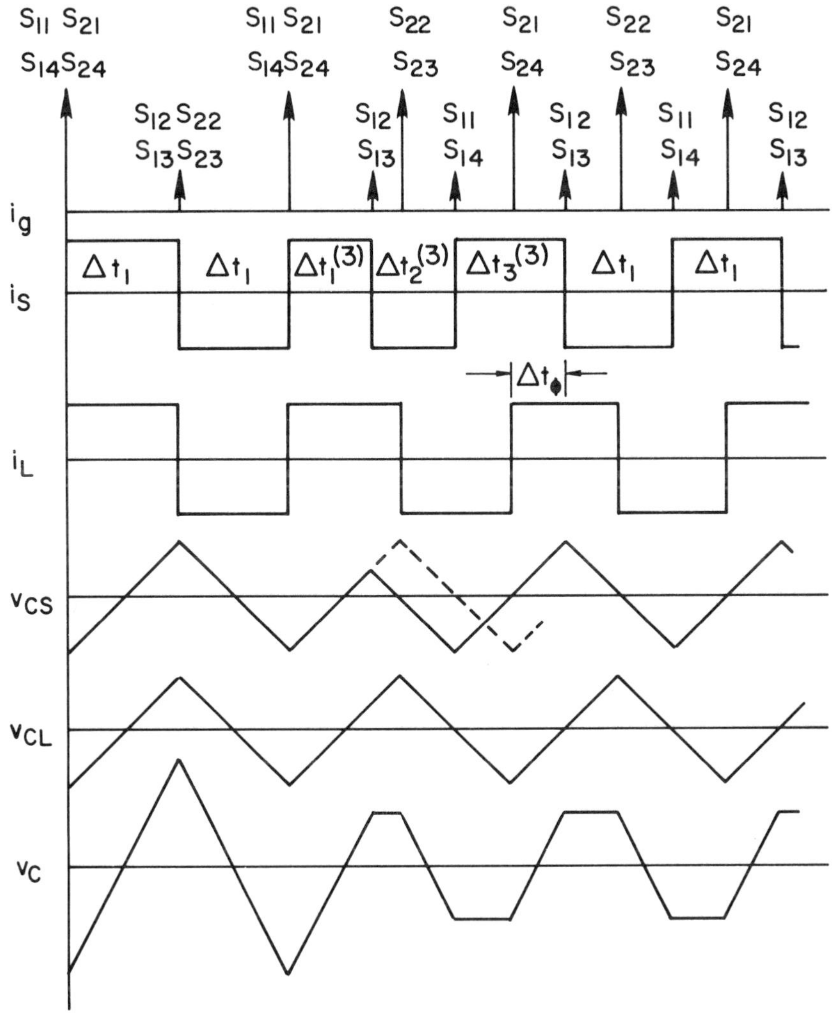

FIG. 6.4. Three-step transient-sequence phase shifting of $\phi = 0° \rightarrow -90°$ at $i_L/i_S = 1$ in the one-phase ICB.

safe threshold angles versus i_L/i_S will be derived later in this chapter. The above conclusion is strikingly different from that of the three-phase ICB. The one-phase ICB not only has a simpler operating switching pattern, but also has the much simpler three-step transient for implementation of bang-bang control, whereas the three-phase ICB requires several more steps, as will be shown.

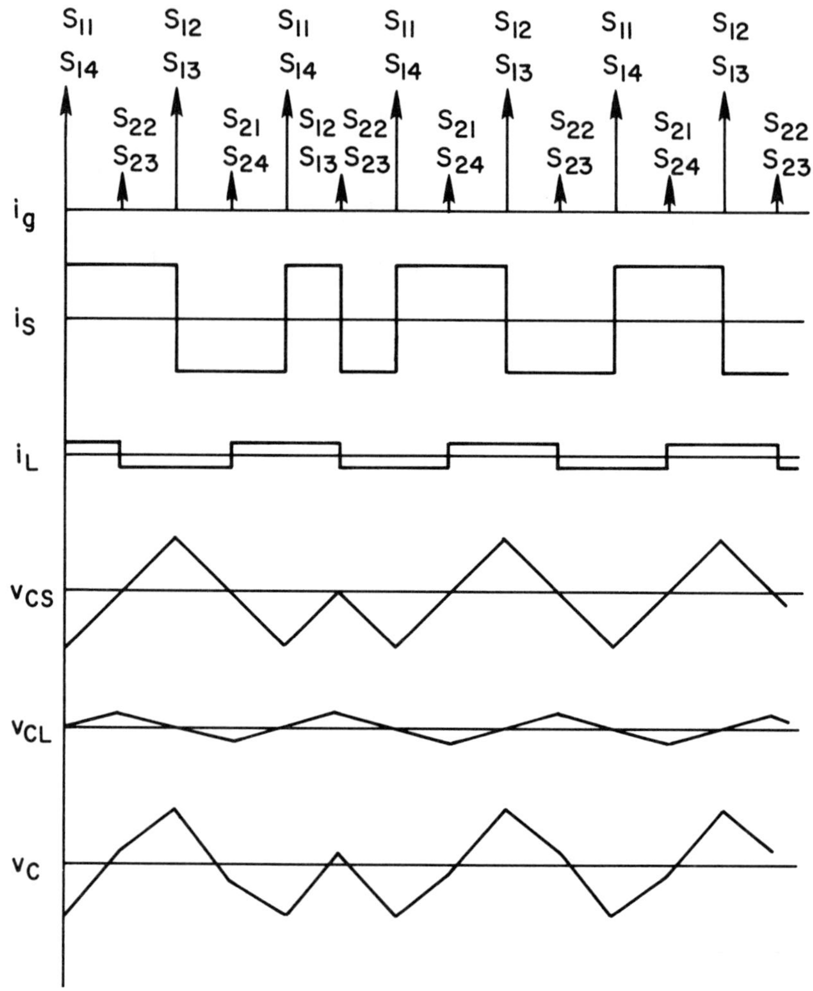

FIG. 6.5. Three-step transient-sequence phase shifting of $\phi = 90° \rightarrow -90°$ at $i_L i_S \simeq 0$ in the one-phase ICB.

For conservative reasons it may be desired to implement the 180° shifts, dictated by a bang-bang algorithm, in two three-step sequences. In this case the two sequences can obviously be implemented in isolation from each other. For example, the one-phase experiments presented in the next chapter use two three-step sequences for $+90°$ to 0° and 0° to $-90°$, separated by one switching interval. However, for optimal time, the bang-bang phase shifts should be implemented as quickly as possible. Two three-step sequences may be merged to share one or two of their steps. In this case a composite five-step or four-step sequence will result. For example, consider two three-step sequences intended to implement Δt_ϕ and $\Delta t_\phi'$ time shifts, respectively:

Sequence 1

$$\Delta t_{11}^{(3)} = \Delta t_1 - \Delta t_\phi/2 \tag{6.4}$$

$$\Delta t_{21}^{(3)} = \Delta t_1 - \Delta t_\phi/2 \tag{6.5}$$

$$\Delta t_{31}^{(3)} = \Delta t_1 \tag{6.6}$$

Sequence 2

$$\Delta t_{12}^{(3)} = \Delta t_1 - \Delta t_\phi'/2 \tag{6.7}$$

$$\Delta t_{22}^{(3)} = \Delta t_1 - \Delta t_\phi'/2 \tag{6.8}$$

$$\Delta t_{32}^{(3)} = \Delta t_1 . \tag{6.9}$$

The composite five-step transient sequence will be

$$\Delta t_1^{(5)} = \Delta t_1 - \Delta t_\phi/2 \tag{6.10}$$

$$\Delta t_2^{(5)} = \Delta t_1 - \Delta t_\phi/2 \tag{6.11}$$

$$\Delta t_3^{(5)} = \Delta t_1' - \Delta t_\phi/2 \tag{6.12}$$

$$\Delta t_4^{(5)} = \Delta t_1 - \Delta t_\phi'/2 \tag{6.13}$$

$$\Delta t_5^{(5)} = \Delta t_1 \tag{6.14}$$

The composite four-step transient sequence will be

$$\Delta t_1^{(4)} = \Delta t_1 - \Delta t_\phi/2 \tag{6.15}$$

$$\Delta t_2^{(4)} = \Delta t_1 - \Delta t_\phi/2 - \Delta t_\phi'/2 \tag{6.16}$$

$$\Delta t_3^{(4)} = \Delta t_1 - \Delta t_\phi/2 \tag{6.17}$$

$$\Delta t_4^{(4)} = \Delta t_1 . \tag{6.18}$$

For bang-bang control, $\Delta t_\phi = \Delta t_\phi' = T/4$ or 90°. Further merging of the two sequences will produce a large phase shift three-step sequence, which was discussed before. Note that if the total resulting phase shift is within the safe phase thresholds at the given i_L/i_S, then all of the above composite sequences will be safe and more conservative in commutation voltage than the three-step sequence.

6.2.1 Frequency Shifting in the One-Phase ICB

The shortest transient switching sequence without lasting voltage imbalance for frequency shifting in a one-phase ICB is a two-step.

If it is desired to shift from switching interval Δt_1 to Δt_2, the two-step sequence will be

$$\Delta t_1^{(2)} = \frac{1}{2}(\Delta t_1 + \Delta t_2) \tag{6.19}$$

$$\Delta t_2^{(2)} = \Delta t_2. \tag{6.20}$$

Figure 6.6 shows an example of a frequency shift upward, and Figure 6.7 shows an example for frequency shift downward. Note that the frequency-shifting transient sequence is implemented on both converters as their switching times arrive. If the operating phase is other than zero, the above frequency shifting will substantially alter the phase relationship after the frequency shift. For example, Figure 6.6 shows operation at $\phi = 45°$ prior to the frequency shift. The phase angle is changed to $\phi = 90°$ following the frequency transient sequence. A similar effect is observed in Figure 6.7, where the phase has been changed inadvertently from $\phi = 90°$ to $\phi = 45°$. Furthermore, the unintentional phase shift is dependent on which converter begins the frequency-shifting transient first. For example, in Figure 6.8 the ICB begins with the same initial phase as in Figure 6.6, $\phi = 45°$, and goes through the same frequency shift. However, in this case the frequency transient sequence happens to begin on the storage converter first, as seen on the v_{CL} and i_L traces. The resulting phase after the transient is sent to $\phi = -90°$.

The unintentional phase shifting associated with frequency shifting can cause two difficulties. First, phase control command is, at least temporarily, distorted. This will reduce the effectiveness of the phase controller. Second, and more importantly, the inadvertent phase shift may send the new phase outside the safe operating phase thresholds for the present i_L/i_S state. Such commutation failures will not occur for the examples cited in Figures 6.6, 6.7, and 6.8 because the initial phases were deliberately chosen to be low and the chosen state of $i_L/i_S = 1$ is the least restrictive. However, if $i_L/i_S \simeq 0$ and $\phi = 90°$ were the case in Figure 6.6, the resulting phase, $\phi = 180°$, would cause a

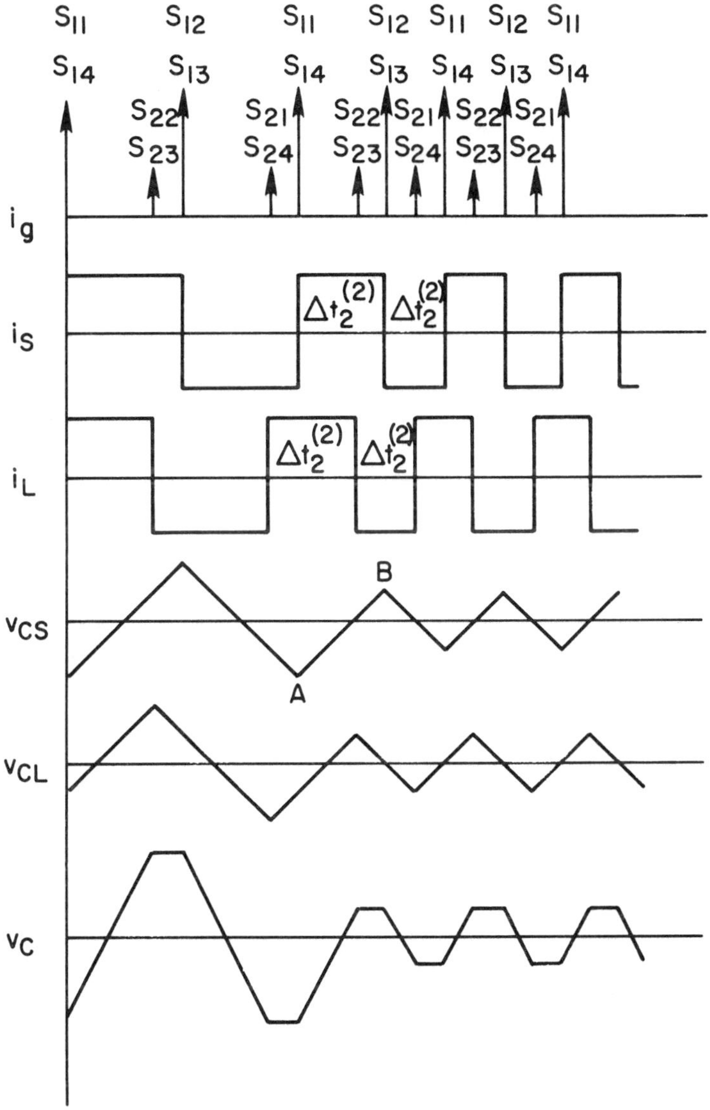

FIG. 6.6. Two-step transient-sequence frequency shifting in the one-phase ICB; frequency increase.

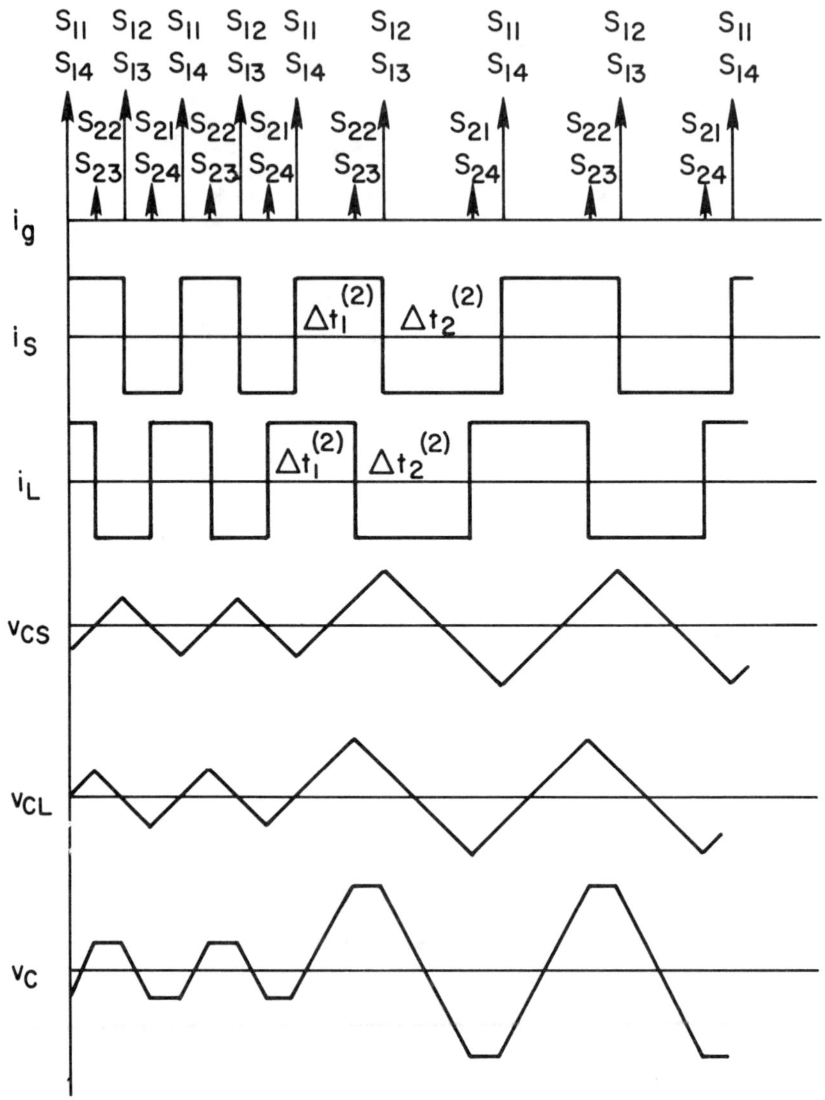

FIG. 6.7. Two-step transient-sequence frequency shifting in the one-phase ICB; frequency decrease.

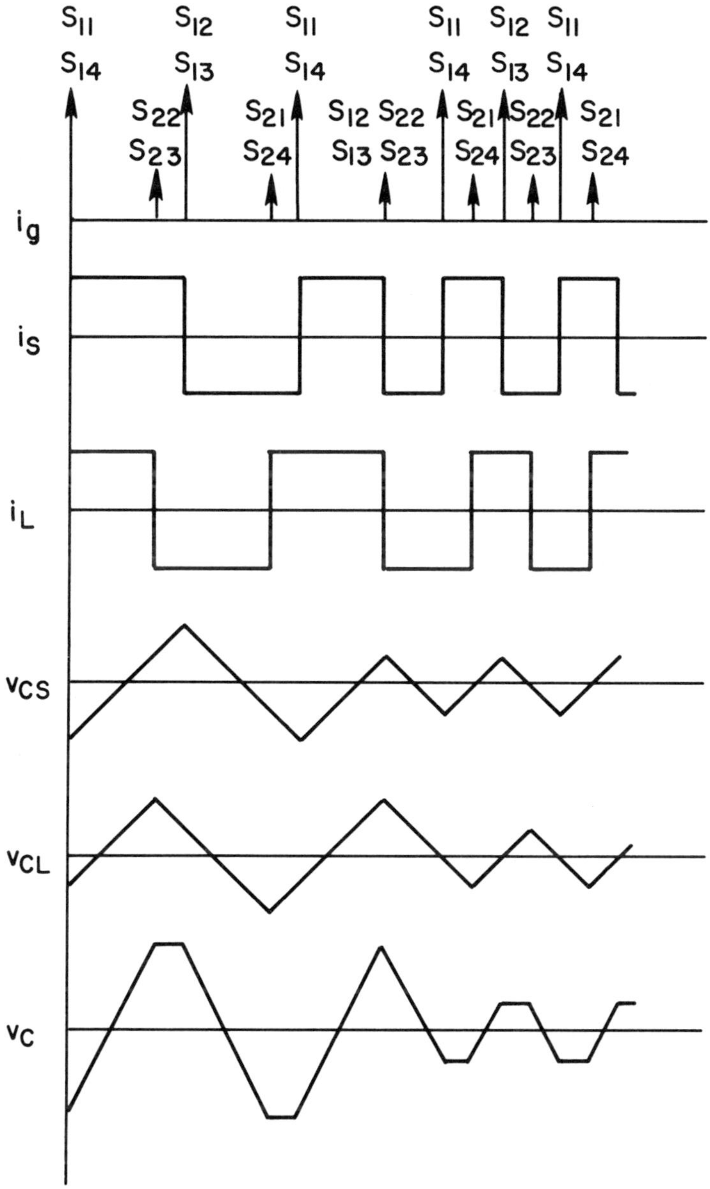

FIG. 6.8. Two-step transient-sequence frequency shifting in the one-phase ICB. Transient sequence began on the lagging converter.

commutation failure. Two techniques can be used to avoid these difficulties. A simple technique is always to phase shift to zero prior to a frequency shift. Zero phase will survive a frequency transient sequence unchanged. Then a phase shift can return the system to the desired operating phase. By a merging of the phase-shifting and frequency-shifting transient sequences by one step, a composite transient sequence will result. Note that the initial phase-shifting transient is implemented on the lagging converter, while the final phase-shifting transient is implemented on the other converter to recover the original phase. Of course, the frequency-shifting transient is implemented on both converters. Therefore, each converter will see a composite four-step transient sequence as follows:

Originally Lagging Converter

$$\Delta t_{11}^{(4)} = \Delta t_1 - \Delta t_\phi/2 \tag{6.21}$$

$$\Delta t_{21}^{(4)} = \Delta t_1 - \Delta t_\phi/2 \tag{6.22}$$

$$\Delta t_{31}^{(4)} = \frac{1}{2}(\Delta t_1 + \Delta t_2) \tag{6.23}$$

$$\Delta t_{41}^{(4)} = \Delta t_2 \tag{6.24}$$

Originally Leading Converter

$$\Delta t_{12}^{(4)} = \frac{1}{2}(\Delta t_1 + \Delta t_2) \tag{6.25}$$

$$\Delta t_{22}^{(4)} = \Delta t_2 - \Delta t_\phi'/2 \tag{6.26}$$

$$\Delta t_{32}^{(4)} = \Delta t_2 - \Delta t_\phi'/2 \tag{6.27}$$

$$\Delta t_{42}^{(4)} = \Delta t_2 \tag{6.28}$$

where: Δt_1 = initial switching interval
Δt_2 = desired or final switching interval
Δt_ϕ = the time shift required to bring the converters in phase
$\Delta t_\phi'$ = the time shift required to implement the original phase at the new frequency

Steps (6.23) and (6.25) are implemented simultaneously on both converters. An example of this procedure is shown in Figure 6.9, in which

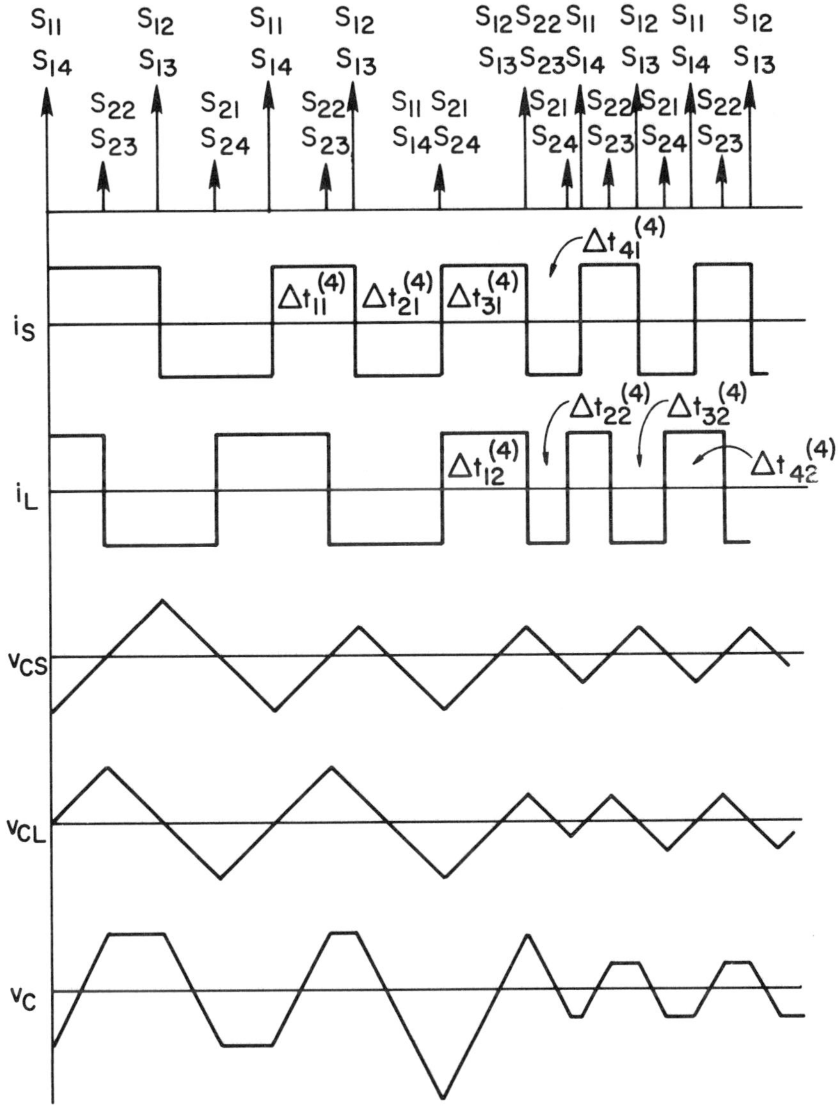

FIG. 6.9. Double four-step composite phase-frequency shifting to maintain constant phase in the one-phase ICB.

a $\phi = 90°$ is first shifted to $\phi = 0°$ and then back to $\phi = 90°$ following the frequency shift. Note that the phase-shift, frequency-shift, phase-shift sequences do not have to be merged and may be implemented as separate sequences. Another technique for avoiding inadvertent phase shifts is to implement a totally merged phase and frequency shifting transient sequence so that as the frequency is being changed, the phase is also being shifted by a correcting amount to compensate for the expected phase error. This technique requires that the frequency-shifting sequence always start on either the leading or the lagging converter so that the phase error can be calculated and compensated for.

6.2.2 Safe Operating Limits in the One-Phase ICB

As has been mentioned previously, the capacitor in the one-phase ICB provides the voltage polarities necessary for commutating the SCRs of the two converters. For the current and voltage conventions shown on the circuit of Figure 4.1, commutation from S_1 and S_4 to S_2 and S_3 in each converter requires a positive capacitor voltage and vice versa. Depending on the state of current distribution in the two inductors, represented by i_L/i_S, and the phase difference between the two converters, such capacitor voltage polarities may not be available at the time of commutations. An example of this is shown in Figure 6.10. Here, $i_L/i_S \simeq 0.25$ and a phase shift from $\phi = 0°$ to $\phi = 135°$ is attempted. The switching commands marked by (X) on the i_g trace will fail because of the wrong capacitor voltage polarity. In reality, the current and voltage waveforms will not proceed as shown after the first indicated commutation failure.

It has been mentioned that the capacitor voltage, which determines the success or failure of a commutation event, is itself determined from two current components: the storage converter current i_S, and the load converter current, i_l (Figure 4.1). Brief examination of the capacitor voltage, as in Figure 6.10, will show that SCRs of the converter whose inductor has the larger current will always commutate successfully. Therefore, for the states $i_L/i_S < 1$ only the commutations of the load SCRs, $S_{L1}S_{L4}$ or $S_{L2}S_{L3}$, need to be investigated. Furthermore, the commutation voltages for $S_{L1}S_{L4}$ and $S_{L2}S_{L3}$ are always symmetrical about the zero-volt axis.* Therefore, investigation of only one

*Since v_C is symmetrical about the axis and $S_{L1}S_{L4}$ and $S_{L2}S_{L3}$ are 180° apart from each other, their voltages must be symmetrical (see Figure 6.10).

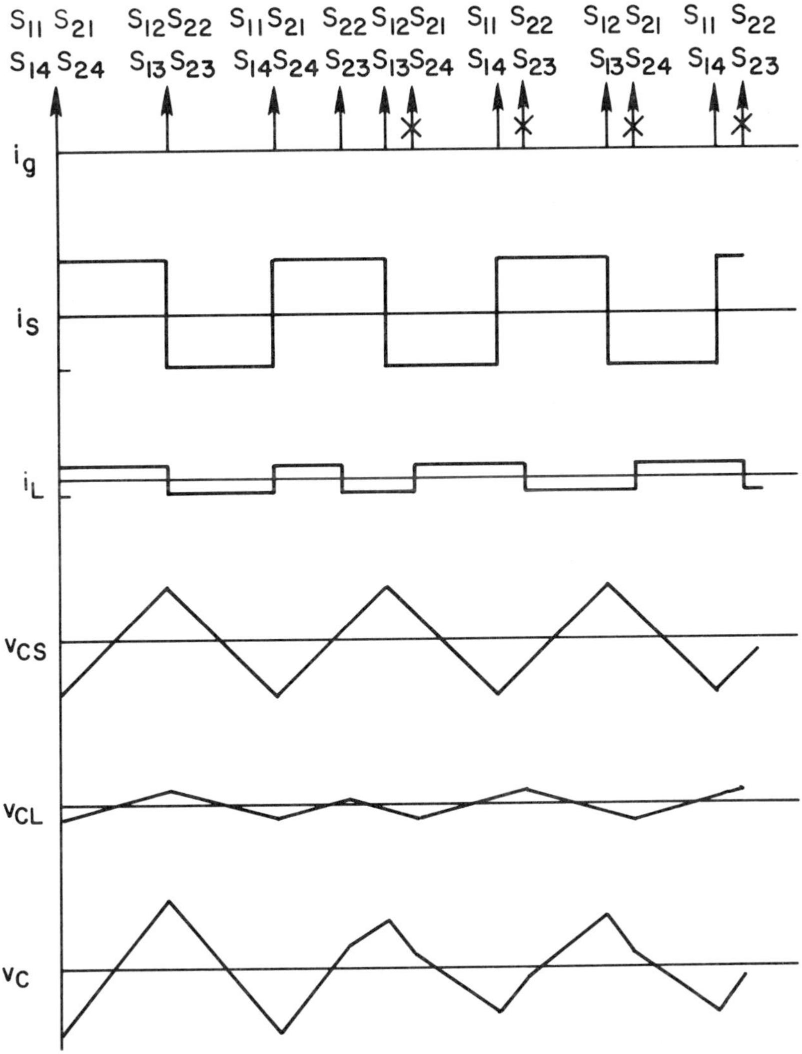

FIG. 6.10. Example of incorrect capacitor commutation voltage resulting from excessive phase difference.

of these voltages will reveal all the commutation outcomes. Investigating the commutation of $S_{L2}S_{L3}$ means checking for the positiveness of the capacitor voltage at the commutation time. The capacitor voltage is

$$v_C = v_{CS} + v_{CL}. \tag{6.29}$$

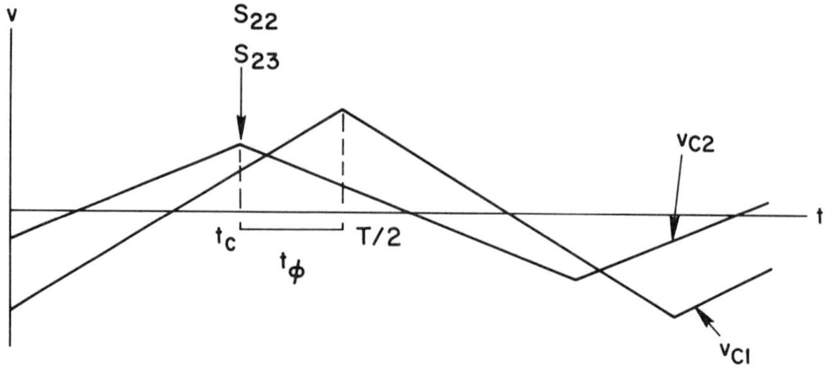

FIG. 6.11. Capacitor component voltages in the vicinity of the $S_{22}S_{23}$ commutation instant in the one-phase ICB.

Figure 6.11 shows an arbitrary case of v_{CS} and v_{CL} when $i_L/i_S < 1$. The values of v_{CS} and v_{CL} at the commutation instant, t_C, can be found from the equations of their first-line segments in Figure 6.11:

$$v_{CS}(t) = \frac{i_S}{C}\left(t - \frac{T}{4}\right) \tag{6.30}$$

$$v_{CL}(t) = \frac{i_L}{C}\left(t - \frac{T}{4} + t_\phi\right) \tag{6.31}$$

$$v_{CS}(t) = v_{CS}\left(\frac{T}{2} - t_\phi\right) = \frac{i_S}{C}\left(\frac{T}{2} - t_\phi - \frac{T}{4}\right)$$

$$= \frac{i_S}{C}\left(\frac{T}{4} - t_\phi\right) \tag{6.32}$$

$$v_{CL}(t_C) = v_{CL}\left(\frac{T}{2} - t_\phi\right) = \frac{i_L}{C}\left(\frac{T}{2} - t_\phi - \frac{T}{4} + t_\phi\right)$$

$$= \frac{i_L T}{4C}. \tag{6.33}$$

The capacitor voltage at the commutation instant is

$$v_C(t_C) = v_{CS}(t_C) + v_{CL}(t_C) = \frac{i_S}{C}\left(\frac{T}{4} - t_\phi\right) + \frac{i_L T}{4C}. \tag{6.34}$$

For successful $S_{L2}S_{L3}$ commutation, we should have

$$v_C > 0 \rightarrow \frac{i_S}{C}\left(\frac{T}{4} - t_\phi\right) + \frac{i_L T}{4C} > 0 \rightarrow \frac{i_L}{i_S} > \left(\frac{4t_\phi}{T} - 1\right). \quad (6.35)$$

After converting t_ϕ to ϕ according to $t_\phi/T = \phi/360°$ and simplifying the result, we will find

$$\frac{\phi}{90} < \frac{i_L}{i_S} + 1. \quad (6.36)$$

A similar calculation can be done for the negative operating phases, $\phi < 0°$, at the states $i_L/i_S < 1$. In this case it is more convenient to investigate the commutation of $S_{L1}S_{L4}$ in Figure 6.12. The task is to check the negativeness of the capacitor voltage at point t_C in the figure. The algebraic problem is identical to the one solved above, except the signs are reversed. The solution for the threshold condition will be

$$\frac{\phi}{90} > -\left(\frac{i_L}{i_S} + 1\right). \quad (6.37)$$

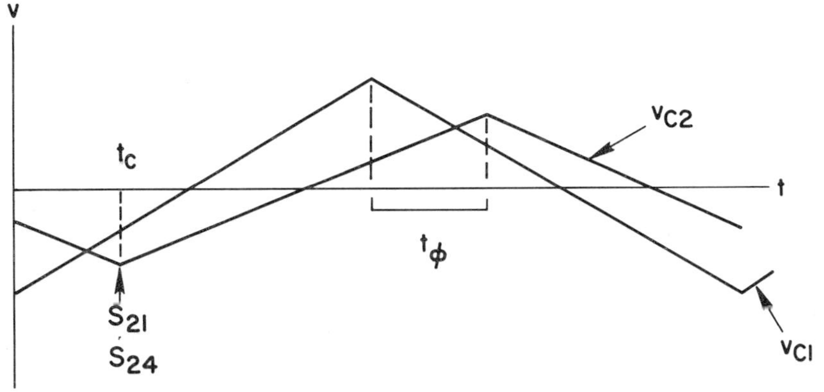

FIG. 6.12. Capacitor component voltages in the vicinity of the $S_{21}S_{24}$ commutation instant in the one-phase ICB.

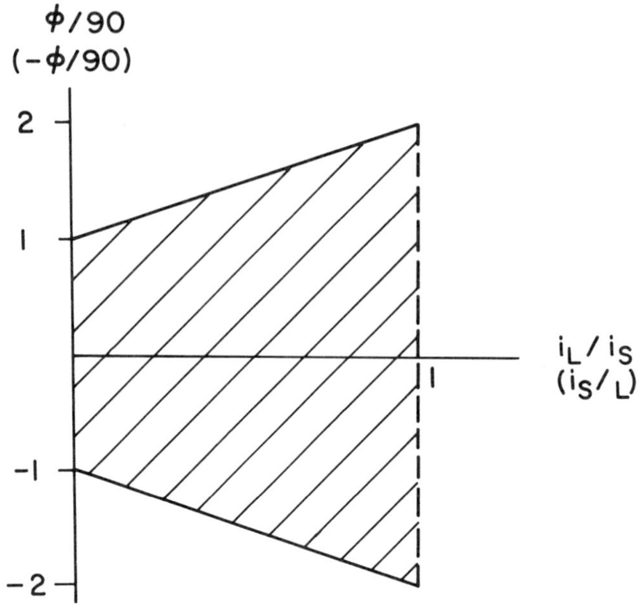

FIG. 6.13. Plot of acceptable operating phase thresholds in one-phase ICB.

A plot of the equations (6.36) and (6.37) threshold lines is shown in Figure 6.13. Operation anywhere within the crosshatched zone in this figure will be possible. Note that the most severe restriction ($|\phi| < 90°$) exists at the beginning of the energy transfer when $i_L/i_S = 0$. This implies that, in theory, bang-bang control strategy will not be able to start the energy transfer because of commutation failures. In reality, bang-bang control will work after a brief transient period, as will be shown in the next chapter. Although the plot of Figure 6.13 gives the phase operating zone up to $i_L/i_S = 1$, ratios of i_L to i_S greater than 1 can be found by reversing subscripts of i_L and i_S and applying reciprocity. For $i_L/i_S = 2$, as an example, one may imagine the roles of the storage and load coils are swapped. Thus, a given ϕ operation at $i_L/i_S = 2$ can be imagined as a $-\phi$ operation at "i_L"/"i_S" $= 0.5$, which is on the graph of Figure 6.13. Therefore, questions on operating phases ϕ at $i_L/i_S > 1$ can be answered by their equivalent questions on operating phases $-\phi$ at $i_S/i_L < 1$, as is shown in parentheses in Figure 6.13. The above discussion indicates that the phases of bang-bang control strategy ($\phi = \pm 90°$) approach the thresholds of safe operation at the beginning and the end of a complete energy transfer cycle.

6.3 Phase Shifting in the Three-Phase ICB

The operating phase of the ICB is altered by delaying or advancing the switching events of one or both converters' sides. The simplest way to retard or advance the switching events and still maintain capacitor voltage balance in a three-phase ICB is a three-step transient switching sequence (Figure 6.14).

In the three-step transient, an identical increment of time is added, subtracted, and added to the next three normal consecutive switching intervals. The prescribed increments can have a positive or a negative algebraic sign, that is, they can be subtracted, added, and subtracted for negative increments. It is shown in Figure 6.14 that the transient switching may be applied to either or both converters, depending on the other constraints. Note, for example, that a storage lead greater than 60° cannot be implemented by three-step transient switching on the storage converter of Figure 6.14(a) alone. The choice between applying the transient switching to the load converter only, for the above example, or to both converters will have to be made on the basis of overvoltage and commutation constraints.

The three-phase ICB requires a minimum of three transient steps for balanced phase alteration. A brief examination of the phase voltage waveforms will show that this is inherent with the three-segmented voltage half-cycle. Other converter switching patterns that may be used for multiphase ICBs would have their own minimum-steps transient based on the number of their half-cycle segments.

As will be shown later, there are circumstances when commutation or other constraints necessitate more than the minimum transient switching steps. Such multistep transients may be produced by superposition of several minimum transients in tandem. For example, a four-step balanced transient in the three-phase ICB may be produced as shown in Figure 6.15. This is also known as the "split-shift" method of phase shifting.

6.3.1 Frequency-Shifting Transients

The frequency of operation of the ICB is altered by increasing or decreasing the time intervals between the switching events of both con-

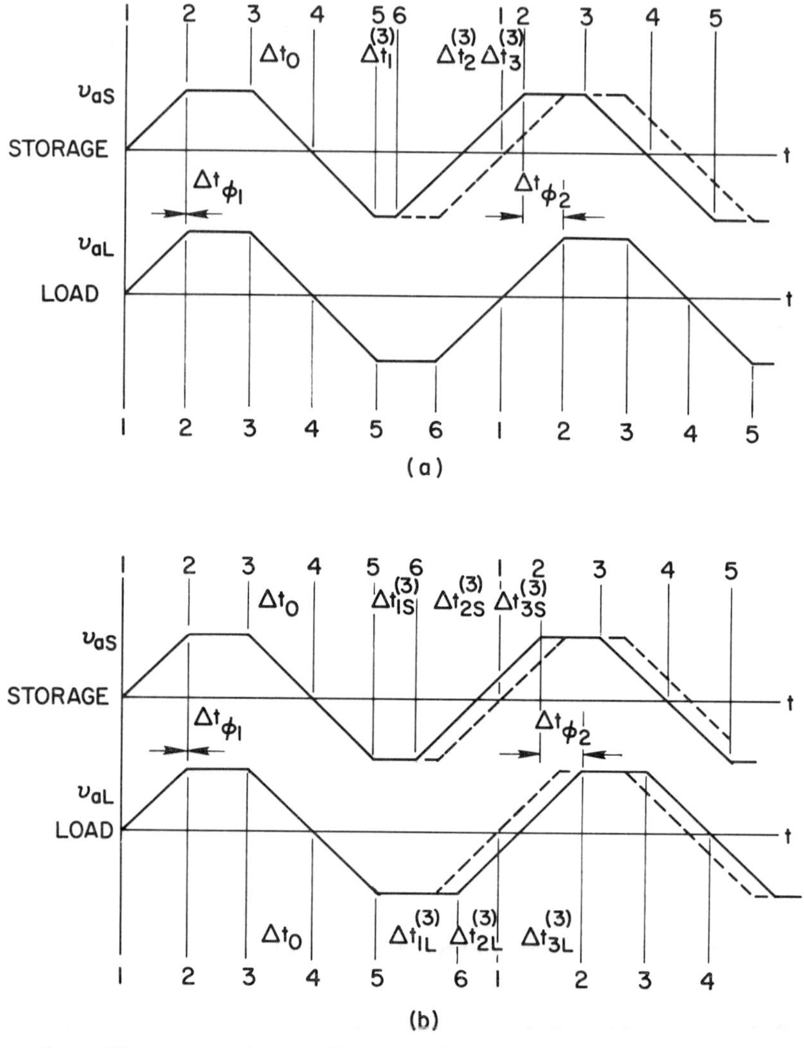

FIG. 6.14. Three-step phase shifting in the three-phase ICB: (a) transient switching on the sorage bridge only; (b) transient switching on both bridges.

verters. It can be shown that for a net power transfer at the fundamental frequency, both converters must operate at the same frequency. The simplest way to change the operating frequency and still maintain balanced phase voltages is the three-step transient switching shown in Figure 6.16. This fundamental three-step transient is an inherent characteristic of the three-phase ICB voltage waveforms. Other switching

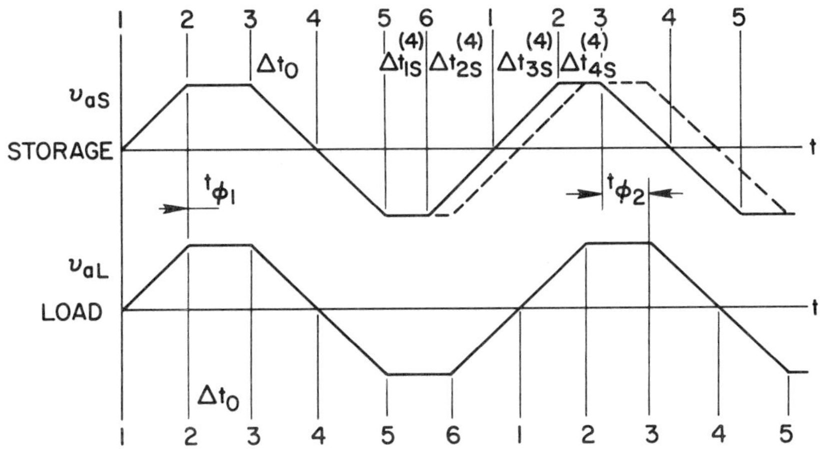

FIG. 6.15. Four-step transient phase shifting in the three-phase ICB.

patterns in multiphase ICBs will have different fundamental frequency-shifting transients.

As was mentioned previously, a pure frequency shift will alter the relative converter phase relationship.* This inadvertent phase alteration may be undesirable. To correct for it a balanced phase-compensating transient must be induced along with the frequency-shifting transient. The shortest total transient would be to superimpose the fundamental phase- and frequency-shifting transients. This would result in a three-step composite transient which can be expressed in matrix form for each converter as

$$S_S^{(3)} = \begin{pmatrix} \Delta t_{1S}^{(3)} \\ \Delta t_{2S}^{(3)} \\ \Delta t_{3S}^{(3)} \end{pmatrix} = \begin{pmatrix} 1 & 1 & 1 \\ 1 & 0 & -1 \\ 1 & 1 & 1 \end{pmatrix} \begin{pmatrix} \Delta t_0 \\ \Delta t_2 \\ \Delta t_{3S} \end{pmatrix} \tag{6.38}$$

$$S_L^{(3)} = \begin{pmatrix} \Delta t_{1L}^{(3)} \\ \Delta t_{2L}^{(3)} \\ \Delta t_{3L}^{(3)} \end{pmatrix} = \begin{pmatrix} 1 & 1 & 1 \\ 1 & 0 & -1 \\ 1 & 1 & 1 \end{pmatrix} \begin{pmatrix} \Delta t_0 \\ \Delta t_2 \\ \Delta t_{3L} \end{pmatrix}, \tag{6.39}$$

where $\Delta t_2 = \Delta t_1 - \Delta t_0 =$ change in the firing intervals due to frequency shift, $\Delta t_{3S} - \Delta t_{sL} =$ the total relative timing compensation to

*If the phase difference before the frequency shifting is an integral multiple of 60°, no phase alteration will occur at the end of the frequency shift.

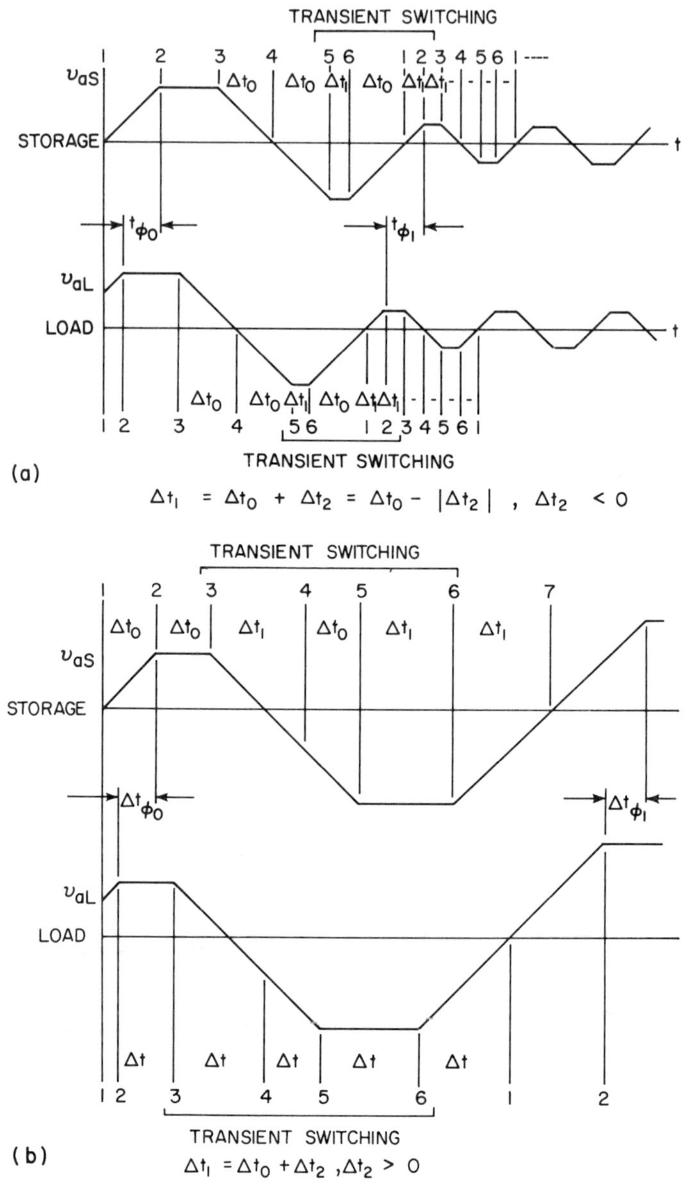

FIG. 6.16. The fundamental three-step frequency-changing transient in the three-phase ICB: (a) frequency increase; (b) frequency decrease.

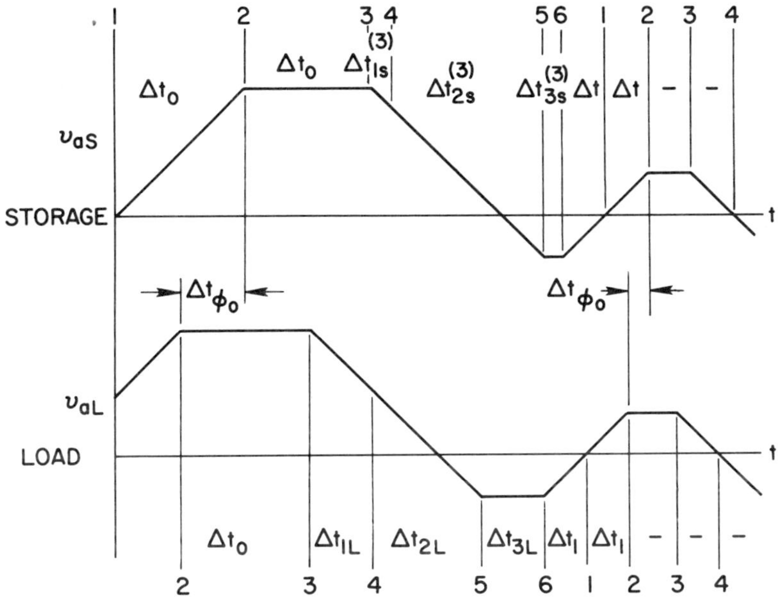

FIG. 6.17. An example of composite three-step transient switching for frequency change while maintaining constant phase in the three-phase ICB.

maintain constant phase, and the rest of the symbols are defined in Appendix 1. Equations (6.38) and (6.39) may also be expressed in the form of individual equations:

$$\begin{aligned}
\Delta t_{1S}^{(3)} &= \Delta t_0 + \Delta t_2 + \Delta t_{3S} \\
\Delta t_{2S}^{(3)} &= \Delta t_0 - \Delta t_{3S} \\
\Delta t_{3S}^{(3)} &= \Delta t_0 + \Delta t_2 + \Delta t_{3S} \\
\Delta t_{1L}^{(3)} &= \Delta t_0 + \Delta t_2 + \Delta t_{3L} \\
\Delta t_{2L}^{(3)} &= \Delta t_0 - \Delta t_{3L} \\
\Delta t_{3L}^{(3)} &= \Delta t_0 + \Delta t_2 + \Delta t_{3L}.
\end{aligned} \tag{6.40}$$

The above intervals are implemented by the SCR triggering circuitry. Figure 6.17 is an example of the implementation of equations (6.40) and (6.41). The values of Δt_S and Δt_L are calculated from the circuit constraints equations,* and the value of Δt_2 is found from the new

* In Figure 6.17, these values are arbitrarily chosen as equal.

frequency specified by the controlling system. If more constraints are imposed than can be satisfied by the three-step transient parameters, Δt_{3S} and Δt_{3L}, then more transient steps must be considered.

From the above discussion it is obvious that phase and frequency can be controlled simultaneously in the same transient switching sequence. For phase-frequency modulation, the mentioned compensating phase-shift transient is adjusted to meet the new phase requirement.

6.3.2 The Four-Step Transient in the Three-Phase ICB

The change of three-phase ICB operating frequency in four transient steps was the first method devised by the authors. The four-step transient is described in some detail here because of its unique advantages. The matrix representations of the four-step transient sequence for the storage and load converters are, respectively,

$$S_S^{(4)} = \begin{pmatrix} \Delta t_{1S}^{(4)} \\ \Delta t_{2S}^{(4)} \\ \Delta t_{3S}^{(4)} \\ \Delta t_{4S}^{(4)} \end{pmatrix} = \begin{pmatrix} 1 & 1 & 1 & 0 \\ 1 & 0 & -1 & 1 \\ 1 & 1 & 1 & -1 \\ 1 & 1 & 0 & 1 \end{pmatrix} \begin{pmatrix} \Delta t_0 \\ \Delta t_2 \\ \Delta t_{3S} \\ \Delta t_{4S} \end{pmatrix} \tag{6.42}$$

$$S_L^{(4)} = \begin{pmatrix} \Delta t_{1L}^{(4)} \\ \Delta t_{2L}^{(4)} \\ \Delta t_{3L}^{(4)} \\ \Delta t_{4L}^{(4)} \end{pmatrix} = \begin{pmatrix} 1 & 1 & 1 & 0 \\ 1 & 0 & -1 & 1 \\ 1 & 1 & 1 & -1 \\ 1 & 1 & 0 & 1 \end{pmatrix} \begin{pmatrix} \Delta t_0 \\ \Delta t_2 \\ \Delta t_{3L} \\ \Delta t_{4L} \end{pmatrix} \tag{6.43}$$

If no frequency shifting is required, $\Delta t_2 = 0$. If it is chosen to induce switching transients on one converter only, e.g., the load converter, equation (6.42) simplifies to

$$e_S^{(4)} = \begin{pmatrix} \Delta t_{1S}^{(4)} \\ \Delta t_{2S}^{(4)} \\ \Delta t_{3S}^{(4)} \\ \Delta t_{4S}^{(4)} \end{pmatrix} = \begin{pmatrix} 1 & 1 & 1 & 0 \\ 1 & 0 & -1 & 1 \\ 1 & 1 & 1 & -1 \\ 1 & 1 & 0 & 1 \end{pmatrix} \begin{pmatrix} \Delta t_0 \\ 0 \\ 0 \\ 0 \end{pmatrix} \tag{6.44}$$

Therefore,

$$\Delta t_{1S}^{(4)} = \Delta t_{2S}^{(4)} = \Delta t_{3S}^{(4)} = \Delta t_{4S}^{(4)} = \Delta t_0,$$

which means steady-state switching. Equation (6.43), with no frequency shifting, will be

$$S_L^{(4)} = \begin{pmatrix} \Delta t_{1L}^{(4)} \\ \Delta t_{2L}^{(4)} \\ \Delta t_{3L}^{(4)} \\ \Delta t_{4L}^{(4)} \end{pmatrix} = \begin{pmatrix} 1 & 1 & 1 & 0 \\ 1 & 0 & -1 & 1 \\ 1 & 1 & 1 & 1 \\ 1 & 1 & 0 & 1 \end{pmatrix} \begin{pmatrix} \Delta t_0 \\ 0 \\ \Delta t_{4L} \\ \Delta t_{4L} \end{pmatrix} \tag{6.45}$$

Furthermore, if each of the two three-step increments imbedded in the above four steps is assigned to execute half of the desired phase shift:

$$\Delta t_{3L} = \Delta t_{4L} = \Delta(\Delta t\phi)/2,$$

the resulting switching intervals are

$$\begin{aligned} \Delta t_{1L}^{(4)} &= \Delta t_0 + \Delta(\Delta t\phi)/2 \\ \Delta t_{2L}^{(4)} &= \Delta t_0 \\ \Delta t_{3L}^{(4)} &= \Delta t_0 \\ \Delta t_{4L}^{(4)} &= \Delta t_0 + \Delta(\Delta t\phi)/2. \end{aligned} \tag{6.46}$$

This is the split-shift method, mentioned previously. Note that in arriving at the special case of the split-shift transient, three assumptions were made: (1) choice of four-step transient; (2) choice of transient on one converter only; (3) choice of equal time increments. These assumptions were made to arrive at any easily implementable phase shifting that will not cause overvoltage or commutation failures. However, it is not always the optimal transient.

The following general observations yield insight into the relative merits of the three-step and four-step transients:

1. For phase shifts of less than 60°, the three-step transient is Δt_0 seconds shorter than the four-step transient.

2. For phase shifts of greater than 60°, the four-step transient is $\Delta t_0/2$ seconds shorter than the three-step transient. This is because the three-step transient can only implement up to 60° phase shift. Therefore, two three-steps are required for phase shifts of greater than 60°.

3. As the desired phase shift is increased to 90°, the four-step transient becomes progressively faster than the three-step. At $\Delta\phi = 90°$, the four-step transient is Δt_0 seconds shorter than the three-step.

4. For both the three-step and the four-step, the switching transient duration is dependent on $\Delta\phi$.

5. The shortest transients are always produced by using negative

increments which reduce the firing intervals on the appropriate bridge side. For example, for $\Delta\phi = 40°$, the three-step increments should be $-40°$, $+40°$, $-40°$,* and the four-step increments should be $-20°$, $0°$, $0°$, $-20°$. Both of these reduce the firing intervals. For $\Delta\phi = 90°$, the three-step increments should be, on the load: $-60°$, $+60°$, $-60°$†; and on the storage: $+30°$, $-30°$, $+30°$. The four-step should be on the load: $-45°$, $0°$, $0°$, $-45°$.

6. The fundamental limitation that makes the three-step transient take longer than the four-step for $\Delta\phi > 60°$, is that a firing interval cannot be reduced by more than 60°. So, the excess required shift has to be taken up by the other converter by positive increments. These positive increments delay the termination time of the switching transient.

7. If converter operation is close to or at the maximum design frequency of the ICB, any negative increment transient will cause commutation failure. Having to use positive increments, the three-step transients will always be faster than the four-step ones in such systems.

8. If converter operation is close to the minimum ICB design frequency, any positive increment transient will cause commutation failure and possibly overvoltages.

6.3.3 General Phase-Frequency Shifting Transients in the Three-Phase ICB

In general an n-step switching sequence, $[S^{(n)}]$; i.e. one which will modify n consecutive switching intervals, may be constructed from the sum of n fundamental switching sequences, $[S_n^{(n)}]$. These fundamental switching sequences are the phase shifting sequence (1, −1, 1 type) and the frequency shifting sequence (1, 0, 1, 1, . . . type). The transient switching sequence, $[S^{(n)}]$, is the sum of n vectors which represent the frequency shifting vector and the phase shifting vectors. The sequences may be defined as

$$[S^{(n)}]^{T\ddagger} \triangleq [\Delta t_1^{(n)}\,\Delta t_2^{(n)}\,\Delta t_3^{(n)}\,\Delta t_4^{(n)}\,\Delta t_5^{(n)}\,\Delta t_6^{(n)}\,\Delta t_7^{(n)}\,\ldots\,\Delta t_n^{(n)}]$$

$$[S_1^{(n)}]^{T} \triangleq \Delta t_0\,[\,1\quad 1\quad 1\quad 1\quad 1\quad 1\quad 1\ldots 1\,]$$

*Or their equivalent in seconds.

†Some commutation margin, e.g., 5°, should be left in practice. This will reduce the usefulness of the three-step transient to ± 55°.

‡Transposition of the vector.

$$
\begin{array}{lllllllllll}
[S_2^{(n)}]^T & \triangleq \Delta t_2 [& 1 & 0 & 1 & 1 & 1 & 1 & 1 \ldots 1 &] \\
[S_3^{(n)}]^T & \triangleq \Delta t_3 [& 1 & -1 & 1 & 0 & 0 & 0 & 0 \ldots 0 &] \\
[S_4^{(n)}]^T & \triangleq \Delta t_4 [& 0 & 1 & -1 & 1 & 0 & 0 & 0 \ldots 0 &] \\
[S_5^{(n)}]^T & \triangleq \Delta t_5 [& 0 & 0 & 1 & -1 & 1 & 0 & 0 \ldots 0 &] \\
[S_6^{(n)}]^T & \triangleq \Delta t_6 [& 0 & 0 & 0 & 1 & -1 & 1 & 0 \ldots 0 &] \\
[S_7^{(n)}]^T & \triangleq \Delta t_7 [& 0 & 0 & 0 & 0 & 1 & -1 & 10 \ldots 1 &] \\
\vdots & & & & & & & & & \\
[S_n^{(n)}]^T & \triangleq \Delta t_n [& 0 & 0 & . & . & . & . & 1 \ -1 \ 1 &]
\end{array}
\quad (6.47)
$$

where $S_2^{(n)}$ is the fundamental (minimum) frequency shifting sequence component to be discussed later.

The symbols used in this analysis are defined in Appendix 1. The n-step sequence vector, $[S^{(n)}]$, consists of n transient switching intervals: $\Delta t_1^{(n)}$, $\Delta t_2^{(n)}$, $\Delta t_3^{(n)}$, . . . $\Delta t_n^{(n)}$ as its elements. Each switching interval, $\Delta t_m^{(n)}$, is the algebraic sum of the steady-state switching interval, Δt_0, and other switching interval increments, such as the incremental interval for frequency shifting, Δt_2, or the incremental intervals belonging to the phase shifting, Δt_3 through Δt_n. In steady-state switching, the transient sequence is identical with the steady-state sequence,

$$S^{(n)} = S_1^{(n)}. \quad (6.48)$$

In the three-step phase shifting sequence, the three-step transient sequence, $[S^{(3)}]$, consists of the sum of the steady-state sequence, and the minimum-phase-shifting incremental sequence, $[S_3^{(n)}|$:

$$S^{(3)} = S_1^{(3)} + S_3^{(3)}, \quad (6.49)$$

where, for simplicity, the brackets have been eliminated. The vector equation (6.49) can be restated in its explicit form as

$$\begin{pmatrix} \Delta t_1^{(3)} \\ \Delta t_2^{(3)} \\ \Delta t_3^{(3)} \end{pmatrix} = \begin{pmatrix} \Delta t_0 \\ \Delta t_0 \\ \Delta t_3 \end{pmatrix} + \begin{pmatrix} \Delta t_3 \\ -\Delta t_3 \\ \Delta t_3 \end{pmatrix} - \begin{pmatrix} \Delta t_0 + \Delta t_3 \\ \Delta t_0 \quad \Delta t_3 \\ \Delta t_0 + \Delta t_3 \end{pmatrix}. \quad (6.50)$$

Equation (6.50) is the mathematical expression of the example of Figure 6.14(a). In that example a storage lead was desired, so the switching increment on the storage converter, Δt_{3S}, was negative.

In general, an n-step transient switching sequence that will alter both the phase and the frequency will be

$$S^{(n)} = S_1^{(n)} + S_2^{(n)} + S_3^{(n)} + \ldots + S_n^{(n)}. \tag{6.51}$$

Again, the vector equation (6.51) may be written in explicit form as

$$\begin{pmatrix} \Delta t_1^{(n)} \\ \Delta t_2^{(n)} \\ \Delta t_3^{(n)} \\ \Delta t_4^{(n)} \\ \Delta t_5^{(n)} \\ \Delta t_6^{(n)} \\ \Delta t_7^{(n)} \\ \vdots \\ \Delta t_n^{(n)} \end{pmatrix} = \begin{pmatrix} \Delta t_0 + \Delta t_2 + \Delta t_3 + 0 + 0 + \ldots \\ \Delta t_0 + 0 - \Delta t_3 + \Delta t_4 + 0 + 0 + \ldots \\ \Delta t_0 + \Delta t_2 + \Delta t_3 - \Delta t_4 + \Delta t_5 + 0 + 0 + \ldots \\ \Delta t_0 + \Delta t_2 + 0 + \Delta t_4 - \Delta t_5 + \Delta t_6 + 0 + 0 + \ldots \\ \Delta t_0 + \Delta t_2 + 0 + 0 + \Delta t_5 - \Delta t_6 + \Delta t_7 + 0 + \ldots \\ \Delta t_0 + \Delta t_2 + 0 + 0 + 0 + \Delta t_6 - \Delta t_7 + \Delta t_8 + 0 + \ldots \\ \Delta t_0 + \Delta t_2 + 0 + 0 + 0 + 0 + \Delta t_7 - \Delta t_8 + \Delta t_9 + 0 + \\ \vdots \\ \Delta t_0 + \Delta t_2 + 0 + 0 + \ldots + \Delta t_n \end{pmatrix}$$

$$= \begin{pmatrix} 1 & 1 & 1 & 0 & \ldots & \ldots & \ldots & \ldots & \ldots & 0 \\ 1 & 0 & -1 & 1 & 0 & \ldots & \ldots & \ldots & \ldots & 0 \\ 1 & 1 & 1 & -1 & 1 & 0 & \ldots & \ldots & \ldots & 0 \\ 1 & 1 & 1 & -1 & 1 & 0 & \ldots & \ldots & \ldots & 0 \\ 1 & 1 & 0 & 0 & 1 & -1 & 1 & 0 & \ldots & 0 \\ \ldots & \ldots & \ldots & \ldots & \ldots & \ldots & \ldots & \ldots & \ldots & \ldots \\ \ldots & \ldots & \ldots & \ldots & \ldots & \ldots & \ldots & \ldots & \ldots & \ldots \\ 1 & 1 & 0 & 0 & \ldots & \ldots & \ldots & \ldots & 0 & 1 \end{pmatrix} \begin{pmatrix} \Delta t_0 \\ \Delta t_2 \\ \Delta t_3 \\ \Delta t_4 \\ \Delta t_5 \\ \Delta t_6 \\ \Delta t_7 \\ \Delta t_n \end{pmatrix}$$

$$= [A_{3\phi}][\Delta t]. \tag{6.52}$$

The topological matrix $[A_{3\phi}]$ may be called the characteristic transient matrix of the three-phase ICB. Characteristic transient matrices may similarly be found for other multiphase ICBs.

As another example, the three-step transient switching of Figure 5.19(b) can be constructed from equation (6.52) as follows: For the storage converter sequence,

$$S_S^{(3)} = \begin{pmatrix} 1 & 1 & 1 \\ 1 & 0 & -1 \\ 1 & 1 & 1 \end{pmatrix} \begin{pmatrix} \Delta t_0 \\ \Delta t_2 \\ \Delta t_{3S} \end{pmatrix}$$

$$= \text{storage converter sequence}. \tag{6.53}$$

Since no frequency shifting is done, $\Delta t_2 = 0$. Since phase shifting to storage lead is done, $\Delta t_{2S} < 0$. Therefore,

$$S_S^{(3)} = \begin{pmatrix} \Delta t_{1S}^{(3)} \\ \Delta t_{2S}^{(3)} \\ \Delta t_{3S}^{(3)} \end{pmatrix} = \begin{pmatrix} 1 & 1 & 1 \\ 1 & 0 & -1 \\ 1 & 1 & 1 \end{pmatrix} \begin{pmatrix} \Delta t_0 \\ 0 \\ -|\Delta t_{3S}| \end{pmatrix} \tag{6.54}$$

Or,

$$\begin{aligned} \Delta t_{1S}^{(3)} &= \Delta t_0 - |\Delta t_{3S}| \\ \Delta t_{2S}^{(3)} &= \Delta t_0 + |\Delta t_{3S}| \\ \Delta t_{3S}^{(3)} &= \Delta t_0 - |\Delta t_{3S}|. \end{aligned} \tag{6.55}$$

For the load converter,

$$S_L^{(3)} = \begin{pmatrix} 1 & 1 & 1 \\ 1 & 0 & -1 \\ 1 & 1 & 1 \end{pmatrix} \begin{pmatrix} \Delta t_0 \\ \Delta t_2 \\ \Delta t_{3L} \end{pmatrix} = \text{load converter sequence.} \tag{6.56}$$

Since no frequency shifting is done, $\Delta t_2 = 0$. Since phase shifting to load lag is done, $\Delta t_{3L} > 0$,

$$S_L^{(3)} = \begin{pmatrix} \Delta t_{1L}^{(3)} \\ \Delta t_{2L}^{(3)} \\ \Delta t_{3L}^{(3)} \end{pmatrix} = \begin{pmatrix} 1 & 1 & 1 \\ 1 & 0 & -1 \\ 1 & 1 & 1 \end{pmatrix} \begin{pmatrix} \Delta t_0 \\ 0 \\ \Delta t_{3L} \end{pmatrix} \tag{6.57}$$

Or,

$$\begin{cases} \Delta t_{1L}^{(3)} = \Delta t_0 + \Delta t_{3L} \\ \Delta t_{2L}^{(3)} = \Delta t_0 - \Delta t_{3L}\,. \\ \Delta t_{3L}^{(3)} = \Delta t_0 + \Delta t_{3L} \end{cases} \tag{6.58}$$

Equations (6.56) and (6.57) are the mathematical expressions of the transients of Figure 6.14(b).

In a typical application, the required phase (time) shift $\Delta(\Delta t\phi)$ is determined by the controlling system.* The unknown increments Δt_{3S} and Δt_{3L} are determined, in part, from the phase constraint equation:

$$\Delta t_{3S} - \Delta t_{3L} = \Delta(\Delta t_{\phi}) = \text{required time shift.} \tag{6.59}$$

The other equation, required to determine the increments, should come from the other five ICB constraints mentioned previously. Once the unknown increments have been determined, the transient switching intervals can be computed from equations (6.54) and (6.57). These intervals are then implemented by the SCR trigger timing circuitry before the steady-state switching is resumed.

It may happen that the other constraints will produce too many nontrivial equations to be satisfied by Δt_{3L} and Δt_{3L}.† In that case more steps should be included in the transient sequence. This will give more degrees of freedom; e.g., Δt_{3S}, Δt_{3L}, Δt_{4S}, Δt_{4L}, to satisfy the constraint equations. If, on the other hand, equation (6.59) is the only nontrivial constraint, then one of the unknowns, e.g., Δt_{3L}, may be assumed equal to zero. This would mean that only the storage converter experiences a transient switching sequence to achieve the new phase relationship.

The general constraint equations (or algorithms) that will express all of the aforementioned constraint statements mathematically (or logically) have not been worked out thus far. The four-step transient, although not optimal, was found to be always adequate for the purposes of phase-modulated control of the prototype ICB system. Therefore, effort was concentrated on developing the control strategy that produces the time shifts $\Delta(\Delta t\phi)$. The present analysis is to be expanded as the need for optimizing multistep transients arises in future ICB systems. These equations, in conjunction with equation (6.52), can provide an analytical way of finding the optimum transient switching sequence for any phase-frequency demand.

6.3.4 The Constraint Equations for Switching Transients in the Three-Phase ICB

It has been mentioned that the required timing difference between the converters is accomplished at the end of the transient switching.

*$(\Delta t\phi)$ is the same as t_0 in Equation (5.69). It has been renamed to avoid confusion with other symbols in this section. See Appendix I.

†The equations may be expressly mathematical or they may be logical (algorithmic).

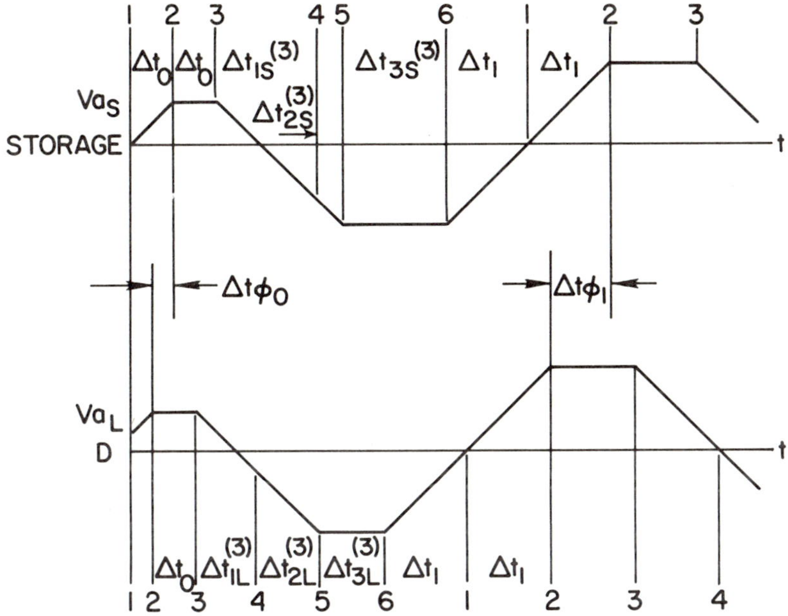

FIG. 6.18. Three-step phase- and frequency-changing transient switching in the three-phase ICB.

The change in timing can be calculated by summing up the transient intervals on each converter, then subtracting the results. This will indicate which converter is left with a net gain in timing. For example, in Figure 6.18, starting from the switching event 3 on each converter, the storage and load converters accumulate the following intervals:

$$\text{storage} : \Delta t_{1S}^{(3)} + \Delta t_{2S}^{(3)} + \Delta t_{3S}^{(3)}$$

$$\text{load} : \Delta t_{2L}^{(3)} + \Delta t_{3L}^{(3)}.$$

These intervals also advance the switching events on each converter to switching event 6. The net timing gain (phase shift) can be calculated as:

$$\Delta[\Delta t\phi] = [\Delta t_{1S}^{(3)} + \Delta t_{2S}^{(3)} + \Delta t_{3S}^{(3)}] - [\Delta t_{1L}^{(3)} + \Delta t_{2L}^{(3)} + \Delta t_{3L}^{(3)}], \quad (6.60)$$

where, as usual, positive timing difference means the load leading. After substitution from equations (6.40) and (6.41),

$$
\begin{aligned}
\Delta[\Delta t\phi] = {} & [(\Delta t_0 + \Delta t_2 + \Delta t_{3S}) + (\Delta t_0 - \Delta t_{3S}) \\
& + (\Delta t_0 + \Delta t_2 + \Delta t_{3S})] - [(\Delta t_0 + \Delta t_2 + \Delta t_{3L}) \\
& + (\Delta t_0 - \Delta t_{3L}) + (\Delta t_0 + \Delta t_2 + \Delta t_{3L})] \\
= {} & \Delta t_{3S} - \Delta t_{3L}.
\end{aligned} \tag{6.61}
$$

This time shift has been added to the original timing difference in Figure 6.18:

$$
\Delta t_{\phi 1} = \Delta t_{\phi 0} + \Delta(\Delta t_\phi). \tag{6.62}
$$

In general form, the phase shift constraint in n-step transient switching is

$$
\begin{aligned}
\Delta(\Delta t_\phi) &= \sum_{n=1}^{n} (\Delta t_{iS}^{(n)} - \Delta t_{iL}^{(n)}) \\
&= \Delta t_{\phi 1} - \Delta t_{\phi 0}.
\end{aligned} \tag{6.63}
$$

If the required phase shift is expressed in degrees at the new frequency, then

$$
\begin{aligned}
\Delta\phi &= \frac{60^\circ}{\Delta t_1} \sum_{i=1}^{n} (\Delta t_{iS}^{(n)} - \Delta t_{iL}^{(n)}) \\
&= \phi_1 - \phi_0.
\end{aligned} \tag{6.64}
$$

If equation (6.64) is expanded by using equation (6.52),

$$
\begin{aligned}
\Delta\phi = \frac{60^\circ}{\Delta t_1} \sum_{i=3}^{n} & [n\Delta t_0 + (n - 1)\Delta t_2 + \Delta t_{iS}] \\
& - [n\Delta t_0 - (n - 1)\Delta t_2 + \Delta t_{iL}.
\end{aligned} \tag{6.65}
$$

After cancellation of the redundant terms,

$$
\begin{aligned}
\Delta\phi &= \frac{60^\circ}{\Delta t_1} \sum_{i=3}^{n} (\Delta t_{iS} - \Delta t_{iL}) \\
&= \phi_1 - \phi_0.
\end{aligned} \tag{6.66}
$$

The constraint equation due to a required frequency shift is, simply

$$\Delta t_2 = \Delta t_1 - \Delta t_0$$
$$= \frac{1}{6}(T_1 - T_0), \tag{6.67}$$

or, in terms of the angular frequency of the converters,

$$\Delta t_2 = \frac{\pi}{3}\left(\frac{1}{\omega_1} - \frac{1}{\omega_0}\right). \tag{6.68}$$

In operation, the controller specifies the new phase, ϕ_1, and the new frequency, ω_1. The initial phase, ϕ_0, and frequency, ω_0, are known. Equations (6.52), (6.66), and (6.68) represent, respectively, constraints 5, 1, and 2 of section 6.1. Constraint 6 can be converted to an iterative algorithm which will try to satisfy the other constraints first by a three-step transient, then by a four-step transient, etc., until the minimum steps of transients are established.* Constraint 3 is an algorithm that limits the maximum interval length, based on the coil currents (see equation [5.15]). It can be implemented as a subroutine of the algorithm of constraint 6. In this capacity, it will prevent the constraint 6 algorithm from exploring transients that cause overvoltage.

Constraint 4 has two implications. The switching intervals should not be reduced to less than the turn-off time rating of the SCRs. In addition, the commutation voltage conditions should at no point be violated. The turn-off time constraint can be implemented as a subroutine of the constraint 6 algorithm, which will negate too short an interval. The commutation voltage constraint either may be formulated in general form or may be verified for a few transient sequence cases of interest. These sequences are then the only options available to the constraint 6 algorithm to choose from.

In an ICB system with a slow-responding load coil, constraint 6

*The algorithm can be further optimized by always attempting to put all the transients on the converter that requires interval shortening. This has been used in the control experiments described later. However, exclusive interval shortening may not be possible for very high frequency ICBs because of violation of SCR turn-off time.

algorithm may be more academic than of practical benefit. Such was the case with the prototype ICB, which was used in the experiments described in the next chapter. The applicable constraints were 1, 3, 4, and 5. These constraints were met by always implementing a four-step transient on the bridge that required interval shortening.

6.4 Discussion of the Phase-Frequency Shifting Analysis

The analysis presented in the previous sections provides a methodical way of changing the phase and/or the frequency at which the ICB operates. The general analysis was confined to the three-phase system because the switching pattern and the dynamic behavior change drastically from one multiphase ICB to another. However, by incorporating this change in the characteristic transient matrix of equation (6.47), the method of this analysis may be applied to any ICB.

The analysis is primarily intended to facilitate the choosing of one or a few efficient transient switching sequences that will suffice in implementing all the shifts of interest in the operation. These transient sequences may then be programmed into the SCR triggering system by software or hardware. It is also possible, in principle, to convert the analysis to an algorithm that will search for the optimal transient sequence in real time. Such an algorithm would be beneficial only if a very wide range of phases and frequencies are to be utilized optimally. It is expected that most practical systems are not capable of such a large control domain. The band of possible operating frequencies is severely limited from both ends. The SCR turn-off time establishes the theoretical upper bound of the band. System efficiency will force the designer to keep the cut-off frequency about a decade below this upper bound. The maximum system voltage rating will establish the lower bound of the frequency band. Most of the benefits of changing the frequency can be obtained by operating only at a few predetermined frequencies which are appropriately located in the permissible band.

Reference

Ehsani, M.; R. L. Kustom; R. E. Fuja; and T. J. Barnard. "General Phase Frequency Shifting in the Three-Phase Inductor-Converter Bridge." *8th Symposium of Engineering Problems of Fusion Research.* San Francisco, Calif., November, 1979.

CHAPTER 7

Feedback Control of the Inductor-Converter Bridge

The inductor-converter bridge (ICB) is used for controlling the magnetic field produced by a load coil. To control the magnetic field, the current or average voltage of the load coil should be actively controlled. If the produced magnetic field is to follow an arbitrary trajectory in real time, then a feedback control is necessary for the ICB system. The feedback controller periodically computes the necessary operating phase and frequency of the ICB, based on equations (5.21), (5.22), (5.23), (5.25), and (5.26), which were derived in chapter 5. Thus, the load field or current can be made to follow a given reference signal. The periodic changes of ICB phase and frequency are then implemented through a series of transient converter switching intervals, which were described in chapter 6.

In this chapter some optimal feedback control strategies for the ICB system are described. These control strategies have been tested by numerical simulations and on an experimental ICB system. Some of the test results on the experimental system are presented in the next chapter.

7.1 Bang-Bang Control Strategy

Bang-bang control strategy is a time-optimal feedback control method in the following sense. For a given step change in the reference or command signal, bang-bang control will drive the ICB so that the output variable will intercept the reference variable in the shortest possible time. For example, if the ICB system output is taken as the load-coil current, bang-bang control will provide the fastest interception of the desired current level by the output current. For the remainder of this chapter the control concepts are discussed with the load-coil current assumed as the system output.

Although optimality of the bang-bang control strategy can be established rigorously through the optimal control theory, the following intuitive justification can render more insight. For the shortest output current step response, the load-coil current must have the greatest rate of rise. In equation (5.22), repeated below, the greatest rate of rise is achieved if the control parameter k is maximized:

$$i_L(t) = I_0 \sqrt{\frac{L_S}{L_L}} \sin \frac{k}{\sqrt{L_S L_L}} t \tag{7.1}$$

where

$$k = \sum_{n=1}^{\infty} \frac{4m}{n^3 \pi^2 \omega C} \left[1 - \cos n \left(\frac{2\pi}{m} + t_p \right) \right] \times \sin^2 \frac{n\pi}{m} \sin n\phi . \tag{7.2}$$

For a given m-phase ICB and switching pattern, k is maximized by the frequency, ω, and phase, ϕ. For example, in a one-phase and three-phase ICB, the k parameters are, respectively,

$$k = \frac{1}{\omega C} \left(\phi - \phi^2/\pi \right), \tag{7.3}$$

and

$$\begin{aligned} &\frac{1}{\omega C}\left(2\phi - \frac{3}{2\pi}\phi^2\right), && 0 \leq \phi \leq \frac{\pi}{3} \\ &\frac{1}{\omega C}\left(3\phi - \frac{3}{\pi}\phi^2 - \frac{\pi}{6}\right), && \frac{\pi}{3} \leq \phi \leq \frac{2\pi}{3} \\ &\frac{1}{\omega C}\left(\phi - \frac{3}{2\pi}\phi^2 + \frac{\pi}{2}\right), && \frac{2\pi}{3} \leq \phi \leq \pi . \end{aligned} \tag{7.4}$$

In the above two cases, k is maximized if $\phi = 90°$ and if ω is the lowest practical value. The lowest frequency, ω, is limited by the maximum safe voltage that can be placed on the ICB capacitors during a complete energy transfer cycle. The frequency, ω, is kept constant at this minimum value. Therefore, for the highest rate of rise of output

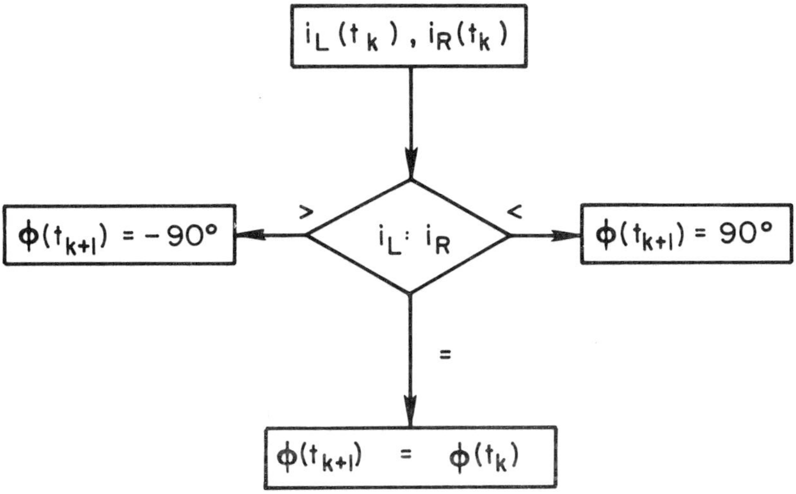

FIG. 7.1. Simple bang-bang control algorithm for one-phase and three-phase ICB systems.

current, the operating phase is $\phi = 90°$. By a similar discussion, based on equations (7.1) through (7.4), it can be shown that the maximum rate of fall of output current can be obtained if ω is kept constant at its minimum value and $\phi = -90°$.

In bang-bang control strategy, therefore, the controller compares the output current with the reference current and makes a choice of the operating phase between those for maximum forward and reverse rate of energy transfer. The frequency is kept constant for this strategy. A flow chart of this simple algorithm for a digital controller is shown in Figure 7.1, where i_L and i_R are the load and the reference currents, respectively.

Operation of the ICB at the phases corresponding to maximum power requires that these phase angles be within the safe commutation thresholds at all times. Figures 4.8 and 6.13 show that $\phi = \pm 90°$ is always within the safe boundaries for the one-phase and three-phase systems. At the beginning of the energy transfer, $i_L/i_S = 0$, the 90° phase angle is at the verge of commutation failure. However, the failing commutations are on the load-side converter, which carries no current at this instant. After some current is established in the load coil, the bang-bang operation is within the safe boundaries.

The bang-bang control algorithm can be programmed in a microcomputer which is also processing the transient and steady-state switching of the ICB. The simplicity of the algorithm significantly reduces the processing time on the microcomputer. This short real-time processing will enhance the accuracy of the control, as shown in the following sections.

7.1.1 Microcomputer-Based Bang-Bang Control

In this section the ICB system is assumed to be a one-phase or a three-phase. In bang-bang control of these systems the operating phase, $\phi(t)$, is shifted from one extreme, $-90°$, to the other, $+90°$, according to the sign of the error quantity, $[i_L(t) - i_R(t)]$ at any given time. The bang-bang control strategy, as described in the last section, renders no decision for the singularity case of $[i_L(t) - i_R(t)] = 0$. Here $i_L(t)$ is the instantaneous average load current and $i_R(t)$ is the instantaneous reference current.

When the error signal is less than one quantization level of the computer, no error is sensed by the control algorithm. Since the quantization level of a finite word-length computer occupies a finite range, a finite interval of time is spent in this singularity, even though $i_L(t)$ and $i_R(t)$ both may be changing (Figure 7.2). Therefore, a separate prescription for commanding the phase during the singularity intervals must be provided.

The simplest approach is to leave the phase unaltered during the singularity intervals. Such a bang-bang control algorithm is shown in Figure 7.1. In this figure the next phase decision, $\phi(t_{k+1})$, is related to the samples of present load current, $i_L(t_k)$, and present reference, $i_R(t_k)$, or, if singular, to the present phase angle, $\phi(t_k)$. The expected system behavior is shown in Figure 7.3.

7.1.2 Accuracy of Bang-Bang Control

The bang-bang control algorithm may be implemented in the microcomputer which is also generating the SCR triggering pulses. In this case the accuracy with which the load current follows a stationary reference signal is a good measure of the overall control system accuracy. This section describes some of the basic concepts that will help predict this accuracy.

The digital controller operates on the quantized version of the load and reference signals. An n-bit analog to a digital converter can work with quantized numbers as large as 2^n. If two's complement tech-

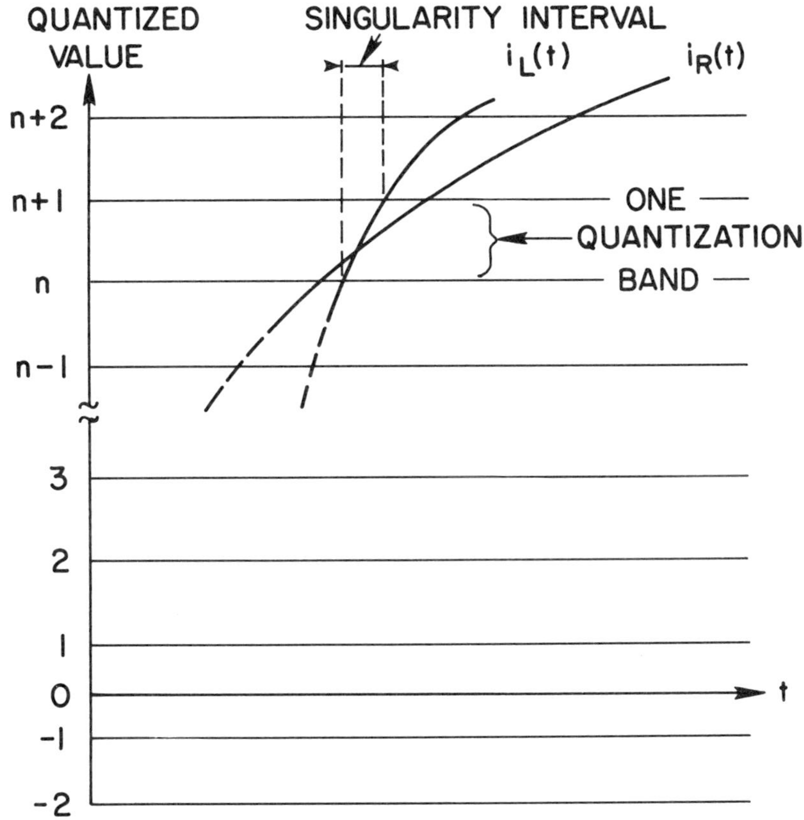

FIG. 7.2. The singularity interval in a finite quantization band.

nique is used to represent negative signals, then the maximum digital number available is 2^{n-1}. If the maximum possible load current is I_{max} A, then this corresponds to the largest number in the microcomputer. In that case the current quantization level of the controller is

$$q = \frac{I_{max}}{2^{n-1}}.$$

The current axis of the controller between 0 A and I_{max} A is divided into 2^{n-1} quantization bands. A stationary reference signal occupies one of these bands, which may be called the reference band. If the controller could respond instantaneously, the flow chart of Figure 7.1 and the waveform of Figure 7.3 show that the load current would os-

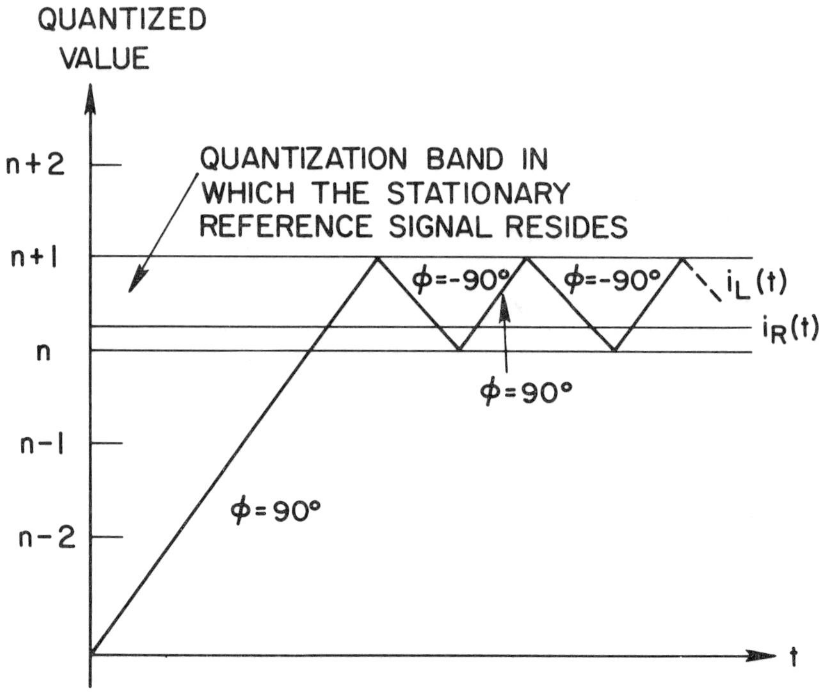

FIG. 7.3. Ideal microcomputer-based bang-bang response to a stationary reference in one-phase and three-phase ICB.

cillate between the thresholds of the reference band. Thus, the theoretical error resulting from quantization may be defined as a maximum of one quantization level, q, or an average of $\pm q/2$. This corresponds to an average percent error of

$$\pm \frac{I_{\max}}{(2^{n-1})I_{\max}} \times 100 = \pm \frac{100}{2^{n-1}}\%.$$

It is evident that in order to reduce the error from quantization, the word length or the number of digits of the controller should be increased.

The actual control system differs from the above idealization in three significant ways. First, the control algorithm, including the phase-shifting software, require a finite processing time. This necessitates signal sampling intervals that allow control algorithm processing to be completed. Second, the ICB system is operating at a finite and modest frequency. Therefore phase shifting between $+90°$ and $-90°$, which requires as many as eight SCR switchings, occupies a non-negligible period of time. Third, the dynamic and static errors introduced by the load current measuring instrumentation must also be considered. The static errors include nonlinearities, ripple, etc. The dynamic errors result from the time-varying signal, such as the transport lag delay of the instrumentation amplifiers. Therefore, to calculate a realistic accuracy, the actual contributions of the above errors must be added to the quantization error.

The instrumentation amplifiers for current probes can be optimally set to behave as a slightly underdamped second-order system. The resonant frequency of such an amplifier is usually set to be much higher than the frequencies present in the load current. It can easily be shown that the steady-state response of such an amplifier to a ramp load current is time-delayed by

$$\Delta t = \frac{2\rho}{\omega_n}, \tag{7.5}$$

where ρ is the amplifier damping ratio and ω_n is the resonant frequency. This causes the amplifier to contribute a transport lag in the control system.

Figure 7.4 visualizes a worst-case instance of load error when all the aforementioned sources of error are accounted for. The last sampling of i_L and i_R that show equality is just at the threshold of the reference band. Thus, the next sampling interval and the software processing time before the instant of phase shifting occur the farthest from the reference band. Then the period with $\phi = 45°$ takes the current still farther from the reference band. Note that while the phase shift is being implemented on the actual i_L, it does not appear on the delayed output of the instrumentation amplifier until Δt seconds later. Similar current error because of the transport lag delay also occurs at the lower side of the reference band when the phase is switched from $-90°$ to $+90°$.

The statistically observed accuracy will be somewhere between the two extremes illustrated by Figures 7.3 and 7.4.

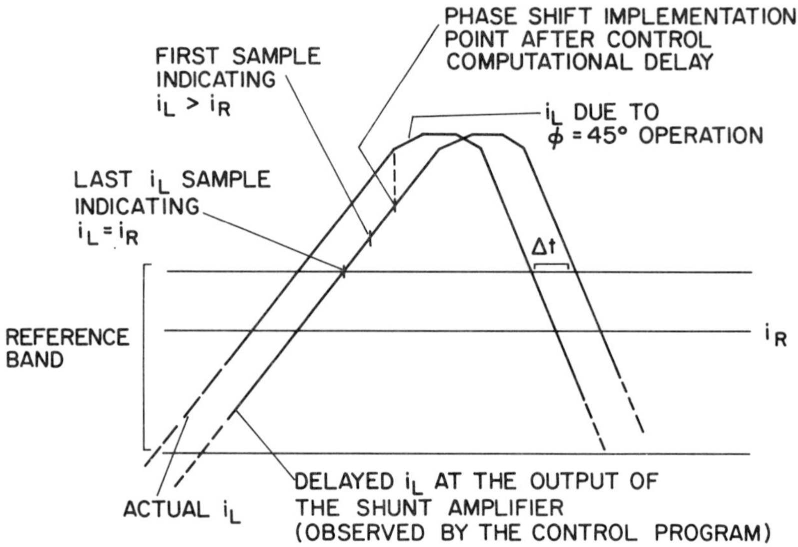

FIG. 7.4. Load current response to control loop delays; worst-case representation. (Not to scale.)

7.1.3 The Effect of Noise in Bang-Bang Control

High-amplitude noise spikes have been observed on the feedback signals of both reference and the load currents. The effect of noise on the amplitude of the bang-bang dither is seen in Figures 7.5 and 7.6. The basic effect of a noise sample of the reference and/or load current is an erroneous bang-bang decision on the basis of the sample pair. Two kinds of erroneous control decisions are possible: an equality decision, resulting in unchanged operating phase, and an inequality decision, resulting in $\pm 90°$ operating phase change. Two erroneous equality cases are visualized in Figure 7.7 in which noise samples on i_L and i_R make the unequal signals appear as equal to the control algorithm. The effect of this false sample pair is the continuation of i_L at $\phi = 90°$ which will increase the response error. Conceptually a sequence of false samples can lead to the load current far away from the stationary reference. However, the probability of consecutive false samples is very small indeed. Moreover, if the load current wanders far enough away from the reference, even the highest amplitude noise sample cannot produce the erroneous equality decision, and the controller will reverse the trend away from absolute instability. It is im-

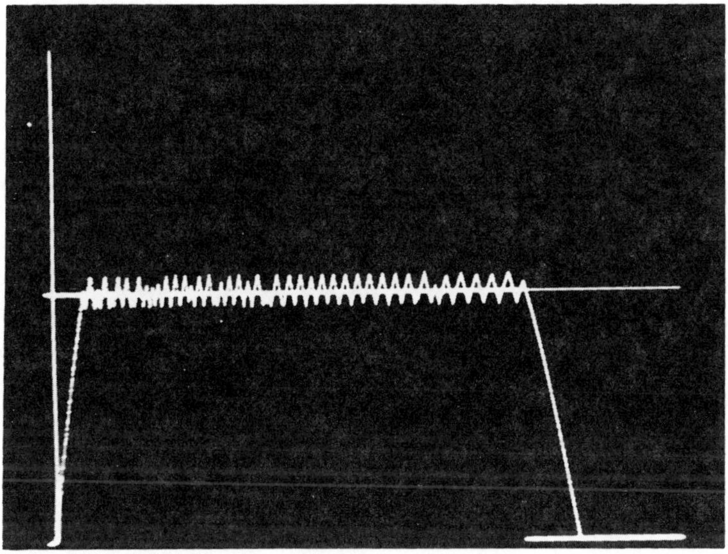

FIG. 7.5. Bang-bang response to a step reference in the presence of noise. The reference trace is superimposed on the i_L (5-bit word length).

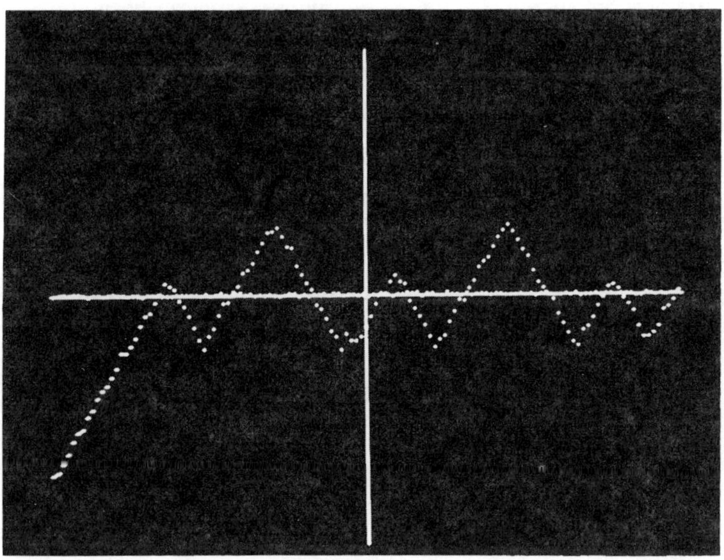

FIG. 7.6. Magnified view of the noise-purturbed i_l dither of Figure 7.5. The reference trace is along the horizontal line.

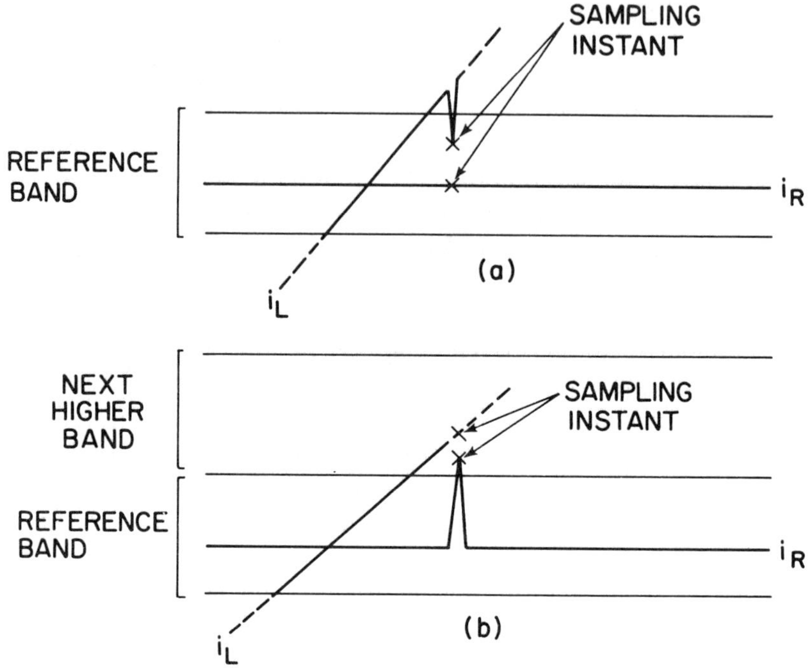

FIG. 7.7. Erroneous equality noise sampling (a) caused by noise on i_L, (b) caused by noise on i_R.

portant to note that false equality samples, which result in response errors larger than the maximum design expectation, have rarely been observed experimentally. This is because the low-frequency dither (less than 36 Hz) spends relatively few samples outside the reference band. Therefore, the probability of intercepting the noise, outside the reference band, is very small, and the chances of consecutive noise samples are even more remote. It is reiterated that false equality control decisions are corrected upon the next true sample pair.

Figure 7.8 visualizes two cases of false inequality samples in which noise samples on i_L and i_R make the digitally equal signals appear as unequal to the control algorithm. The effect of these sample pairs is to change the operating phase according to the bang-bang strategy. Note that since the load current remains within the reference band, which is the singularity zone of the bang-bang algorithm, even the subsequent true samples will not reverse the trend set by the false sample pair until the load current leaves the reference band. However, this erroneous response temporarily reduces the load current error below the design level, as the dither is contained within one quantization band. The penalty for this increased dither frequency is additional switching losses, which will be discussed later. The increase of dither frequency because of noise samples has been frequently observed experimentally. Since the i_L and i_R signals are sampled within the same quantization band, many tens of times per dither period, the probability of a false inequality sample pair is significantly higher than is the case with false equality.

Two important observations may be made on the basis of the above discussion. The bang-bang control strategy is shown to be strongly noise-tolerant. This capability can reduce the cost of instrumentation for the control system. Second, an attempt to increase the design accuracy by shortening the quantization band will also increase the probability of erroneous control decisions from noise samples. However, the probability of enhanced accuracy because of noise will also tend to increase.

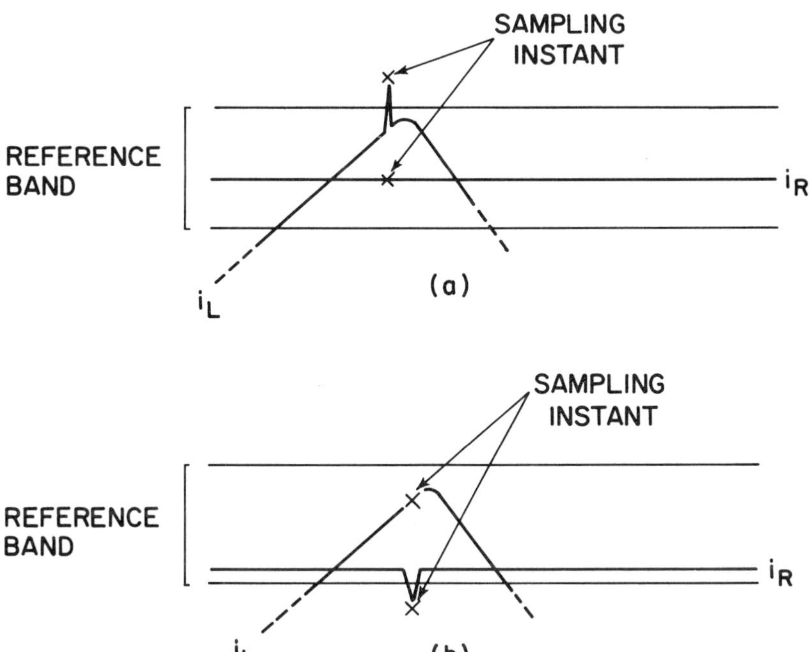

FIG. 7.8. Erroneous inequality noise sampling (a) caused by noise on i_L, (b) caused by noise on i_R.

7.1.4 Switching Losses of Bang-Bang Control

The periodic phase shifting in the bang-bang dither increases the number of SCR switchings in the ICB. To assess the additional converter losses resulting from this control strategy, a worst-case analysis is presented here for the case of a three-phase ICB. The maximum dither frequency occurs when the reference is stationary and the storage coil is fully charged. A brief examination of the double four-step (split-shift) phase shift, required for $\pm 90°$ to $\mp 90°$ transition, can show that this shift adds three additional SCR switchings to one converter cycle. Therefore, a dither frequency of n Hz, adds $3n$ additional switchings per second. Since each converter cycle consists of six switchings, the total uncontrolled system switching per second, at an operating frequency of f Hz, will be

$$m_{\text{no control}} = 12f \text{ switchings/S.} \tag{7.6}$$

The total ICB switchings per second, including the bang-bang dither is then

$$m_{\text{bang-bang}} = 12f + 3n \text{ switchings/S.} \tag{7.7}$$

If the amount of energy dissipated in each SCR switching is assumed to be ΔE, the percentage of the switching loss contributed by the bang-bang control is

$$\begin{aligned} p_{\text{bang-bang}} &= \frac{\Delta E(m_{\text{bang-bang}} - m_{\text{no control}})}{\Delta E(m_{\text{no control}})}(100) \\ &= \frac{3n}{12f}(100) \\ &= \frac{n}{4f}(100\%) \text{ of no-control switching losses.} \end{aligned} \tag{7.8}$$

For the prototype ICB system built at Argonne National Laboratory, the maximum design dither frequency was found to be about 36 Hz at the operating frequency of 631 Hz. Therefore, the bang-bang switching losses were calculated to be

$$\begin{aligned} p_{\text{bang-bang}} &= \frac{36}{4(631)}(100) \\ &\simeq 1.4\% \text{ of no control switching losses.} \end{aligned}$$

This additional loss has a very small impact indeed on the system efficiency. It should be noted that the switching losses themselves constitute only a fraction of the total system losses, which include SCR conduction losses, conductor and contact losses, etc. To put this effect in perspective, in an ICB system with 5% power loss in switching, bang-bang control will increase this to 5.07% switching loss.

The above design efficiency of bang-bang control may be reduced by the increased dither frequency because of noise. However, even if the dither frequency is increased by a factor of two, equation (7.8) shows that the additional losses are still quite small. A theoretical maximum for switching losses resulting from bang-bang control may be calculated as follows. The 180° phase shift required by the bang-bang strategy can be implemented in the course of one converter period. Therefore, the maximum possible dither frequency can be f Hz. Thus, the additional bang-bang losses will be

$$p_{\text{bang-bang}} = \frac{f}{4f}(100) = 25\% \text{ of no-control switching losses.}$$

This additional loss may still be acceptable.

The above discussion is based on the assumption that the phase-shift transients are implemented on the converter that requires shortening of the switching intervals. If the operating frequency is not set at the minimum design level, the phase shifts may be implemented on the opposite converter, which requires interval lengthening. In this case no additional switching losses are introduced at all.

Many refinements of the above loss assessment are possible. For example, if the switching losses in the two converters of the ICB are substantially different, a weighted switching loss may be associated with each converter. The loss calculation may then be carried out separately on the two converters, and the results may be added for the total control losses.

An important conclusion of the discussion of this section is that an attempt to reduce the dither frequency of bang-bang control cannot be justified on the basis of switching losses. The above statement is expected to hold true even for ICB systems with fast-responding load coils.

7.1.5 Discussion of Bang-Bang Control Design

The accuracy of bang-bang control strategy is determined, primarily, by the error present at the time that a sample pair from i_L and i_R show inequality. This error is partly determined by the frequency of sampling. A higher sampling frequency will find inequality between i_L and i_R samples, before i_L has moved too far from i_R. The sampling frequency, in turn, is determined by the speed of real-time processing between the samples. A faster computer will provide shorter sampling intervals to compute the control phase. The simplicity of the bang-bang algorithm is a great asset in developing real-time efficient programs.

For a given sampling rate, the optimal quantization level is one which will always contain about one sampling. In other words, the sampling period should be slightly shorter than the shortest time that i_L takes to cross one quantization band. This determines the size of the processor word length for a given ICB load-current response time. Based on the above discussion, the greatest improvement in the accuracy of control experiments would come from increasing the word length, because more than ten samplings were contained within one band, producing redundant equality decisions, while $(i_L - i_R)$ error was increasing. Note that some improvement in accuracy can be achieved by increasing the sampling rate, because of the computational delay after the inequality sample pair. However, this improvement is almost negligible compared to the quantization error imposed by the limited word length.

The above discussion correlates the four fundamental parameters that influence the accuracy of control: ICB system response, quantization level of the digital processor, processing speed of the digital processor, and the time efficiency of the control software. The user will, in general, specify the ICB system response time and the required accuracy of the load-current control. From the accuracy specification, the control engineer can specify the quantization level, which will set the word length of the control processor. The maximum specified response time of the ICB will determine the shortest quantization-band crossing time by the load current. The sampling period should be set somewhat shorter than this crossing time. From this sampling period, the engineer will decide the required speed of the digital processor and the overall structure of the control software. An efficient control software which has a short cycle time between the sampling and phase angle implementation will require a slower digital processor, whereas, a long and unsophisticated software will require a faster processor to meet the sampling-period constraints. It is apparent that there is a trade-off between the processor hardware and software costs to meet the control specifications. Moreover, all the instrumentation delay is charged against the processing time, as was shown in the accuracy evaluation section. Therefore, high-quality instrumentation may also be traded with processor hardware/software costs. Once a first-round decision has been made on the specifications of the hardware and software, the control engineer will go back and reevaluate the performance of the hypothetical control system. This time, the necessary fine adjustments will be made on the specification of each building block. Finally, a satisfactory and reasonable design will be frozen, and the construction, testing, and evaluation of the total control system will begin.

With the rapid advancement of the computer technology, an increasing amount of the design burden may be put on the processor. For example, a fast computer may be used to oversample i_L and i_R and perform a noise cancellation routine, along with other control routines, and still meet the sampling-period requirement. Thus, the burden of building precision instruments and noise-free analog data links may be reduced. Or, if the sampling-period constraint is difficult to meet, several computers can process consecutive sample pairs in parallel, thus reducing the pressure on any one processor.

7.2 Simultaneous Phase and Frequency Control

Since an ICB is not synchronized to any utility system or otherwise constrained except for component considerations, frequency can be a variable parameter. The switching transients for simultaneous phase and frequency shifting are described in section 6.3.3. Thus, only the control algorithms for the two parameters are needed.

One could, for example, design a system in which the frequency is optimized to operate throughout the energy cycle with the maximum voltage allowed by the components. As was indicated in section 4.2.3, the maximum stress on the capacitors given by equation (4.20) occurs when I_L/I_S is equal to L_S/L_L. For equal inductances, $L_L = L_S$, the maximum voltage occurs when the coil currents are equal. When all of the current is in either one of the coils, the maximum voltage in the system is reduced by $\sqrt{2}$. The frequency, f, could be controlled by the algorithm:

$$f = f_0 I_0/(I_S + I_L),$$

where f_0 is the frequency when all the energy is in one or the other coil.

Other simultaneous phase and frequency control techniques were studied by Hirano, Kustom, and Skiles. These essentially controlled the phase in open-loop control and the frequency in a bang-bang closed-loop control. The primary goal was to reduce the dither frequency of bang-bang control. The details need not and will not be presented here. Basically, an ICB has two parameters that can be controlled in either open- or closed-loop modes to optimize or improve some characteristic performance.

Reference

Ehsani, M., and R. L. Kustom. *Development, Analysis and Control of the Inductor-Converter Bridge*. Argonne National Laboratory Technical Report ANL/FPP/TM-144. Argonne, Ill.: Argonne National Laboratory, 1981.

CHAPTER 8

Some Experimental Results from Model Energy Transfer Systems

In this chapter we will describe some experimental converter circuits for superconductive magnetic energy storage and transfer. Some of the test results and waveforms of these experimental systems are also presented.

8.1 Karlsruhe Single Flying Capacitor Bridge

Dick and Dustmann of the Institut für Technische Physik, Karlsruhe, West Germany, built and tested a model SFC bridge circuit using 10 kJ and 5 kJ superconducting coils as the storage and load coils, respectively. A block diagram of the circuit is shown in Figure 8.1, and the individual component ratings are listed in Table 8.1.

An oscillograph of a typical energy transfer cycle is shown in Figure 8.2 and a listing of the pertinent circuit parameters is presented in Table 8.2.

TABLE 8.1 Parameter List For Karlsruhe 10 kJ Model SFC Bridge Circuit

L_S—Superconducting storage coil with a 17.7 mH inductance, critical current of 1440 A; 22-strand 1,000-filament NbTiCu-matrix wire with 2-mm twist pitch.

L_L—Superconducting load coil with a 17.7 mH inductance and a critical current of 704 A.

P_S—1200 A, 10 V, current-regulated power supply.

C—1–10 parallel units of 384 μF, 400 V metalized paper capacitors.

SCR—4 parallel units of 290 A, 800 V thyristors.

SG—350 V limiters with dump resistors R.

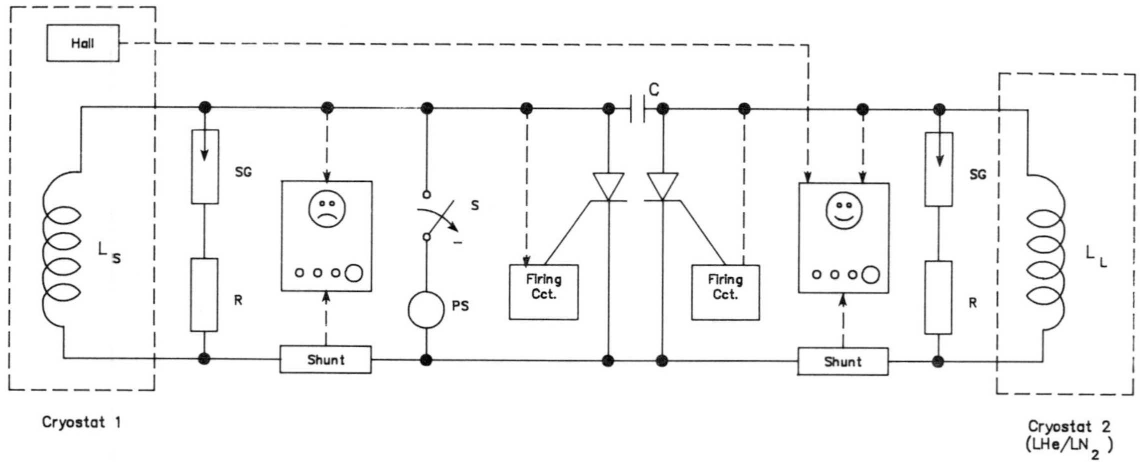

FIG. 8.1. Circuit diagram for Karlsruhe SFC bridge for 10 kJ model coils. (Courtesy of Dr. C. H. Dustmann, Institut für Technische Physik, Karlsruhe, Germany)

FIG. 8.2. Oscillograph of energy transfer cycle with the Karlsruhe SFC bridge. (Courtesy of Dr. C. H. Dustmann, Institut für Technische Physik, Karlsruhe, Germany)

TABLE 8.2 Pertinent Parameters for Energy Transfer Cycle Shown in Figure 8.2

Initial stored energy	5,700 J
Initial storage coil current	800 A
Transfer time	0.45 s
Maximum voltage	130 V
Number of transfer steps	157
Transfer efficiency (% of total energy)	77.1%
Resistive losses (% of total energy)	1.2%
Thyristor losses (% of total energy)	19.0%
Capacitor losses (% of total energy)	1.7%
Superconductive losses (% of total energy)	0.4%
Untransferred energy (% of total energy)	0.6%

8.2 Argonne National Laboratory Three-Phase Inductor-Converter Bridge

Fuja, Kustom, and Smith built an experimental three-phase ICB for 100 kJ coils at the Argonne National Laboratory. A circuit diagram for the ANL bridge is shown in Figure 8.3, and a list of circuit parameters is presented in Table 8.3.

8.2.1 Open-Loop Phase Control Tests

Control by phase modulation required an algorithm for safe implementation of phase shifts. Although the phase-frequency analysis

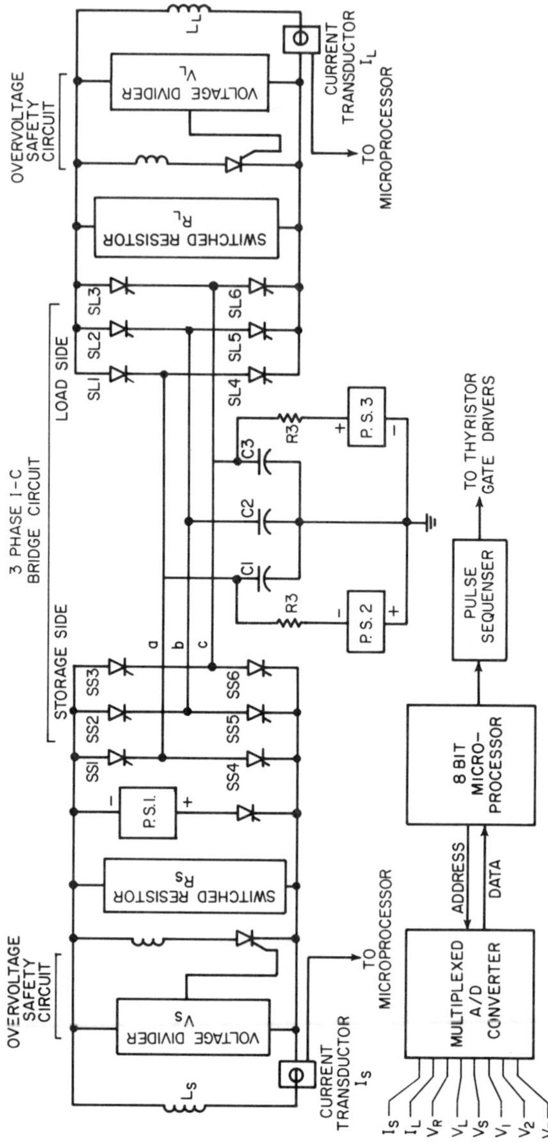

FIG. 8.3. Circuit diagram of ANL three-phase ICB for 100 kJ model coils.

TABLE 8.3 Prototype ICB Specifications

Storage coil inductance (L_S) = 4 H
Load soil inductance (L_L) = 4 H
Angular frequency of converter operation (ω) = 4084 rad/sec
Phase capacitance (C) = 10^{-4}F
Converter period (T) = 1.538 ms
SCR rms current rating (I_T) = 750 A
SCR peak reverse voltage (V_{DRM}) = 500 V
SCR turn-off time (tq) = 20 μs
SCR rate of rise of current (di/dt) = 200 A/μs
SCR rate of rise of recovery voltage (dv/dt) = 800 V/μs
SCR on-state voltage (V_T) = 1.2 V
SCR holding current (I_H) = 800 mA
Initial storage coil current in the experiments (I_0) = 100 A

of chapter 6 indicates that the most efficient switching sequence for phase shifting is the three-step sequence, for the reasons cited in section 6.3.2 the four-step sequence was chosen as the phase-shifting technique. Predominant among these reasons was the inability of the three-step to implement more than a 60° shift in one sequence. Therefore, marginal gains in response time of the three-step, when applicable, were sacrificed for the simplicity of programming only one transient sequence in the control routine for all phase shifts.

A phase-modulating routine that utilized only the four-step transient was developed for the Intel microcomputer. This program accepts a reference command phase and compares it with the operating phase. The difference is then implemented as a phase shift on the appropriate converter. Since in the control experiments the operating frequency was set at the lowest design limit,* any positive increments of the switching intervals could lead to overvoltage failures. Therefore, the phase-shifting routine always implements the required shift on the converter that must receive decrements of switching intervals.

8.2.1.1 Constant-Phase Test. In the initial evaluation experiments, a step voltage was fed to the A/D converter as the reference phase. Fig-

*Minimum frequency corresponds to the maximum power transient and system response time. See equation (5.34).

ure 8.4 shows the average coil voltages and currents in a complete energy transfer from the storage to the load at the phase $\phi = 90°$ or $t_0 = T/4$.

To compare this with the analytical results, the circuit parameters listed in Table 8.3 are substituted into equations (5.21), (5.22), (5.23), (5.25), and (5.26). The value of the k may be calculated from equation (5.34) or in the middle part of equation (5.69), which gives

$$k = \frac{1}{C}\left(3t_0 - \frac{6t_0^2}{T} - \frac{T}{12}\right), \qquad T/6 < t_0 < t/3. \tag{8.1}$$

The results are

$$i_S(t) = 100 \cos 0.5609\, t, \quad A \tag{8.2}$$

$$i_L(t) = 100 \sin 0.5609\, t, \quad A \tag{8.3}$$

$$v_S(t) = 224.4 \sin 0.5609\, t, \quad V \tag{8.4}$$

$$v_L(t) = 224.4 \sin 0.5609\, t, \quad V \tag{8.5}$$

$$\langle p_S \rangle(t) = 11218 \sin 1.1218\, t, \quad W \tag{8.6}$$

A plot of these equations appears in Figure 8.5. Good agreement between the analytical and the experimental results exists. Note, however, that the experimental final load current is somewhat lower than the initial storage current. This is because of the system losses which were not considered in the analysis.*

8.2.1.2 Variable-Phase Tests. To test the capability of the phase-shifting routine in following arbitrary phase commands, $\phi(t)$, input phase signals such as sinusoids, steps, or sawtooths were fed to the microprocessor. Figures 8.6 and 8.7 show the ICB response to such phase functions. The software was calibrated to interpret ± 10 V analog input as ± 90° phase command. In Figure 8.6, the test begins with $\phi = 42°$ and is then stepped to zero and back to 42° for the rest of the

*The shunt resistors in Figure 8.3, which are for initiating SCR holding currents in the coils, were not switched out in these tests. The greater part of the system losses were caused by these resistors. These resistors were switched out by the subsequent versions of the control program.

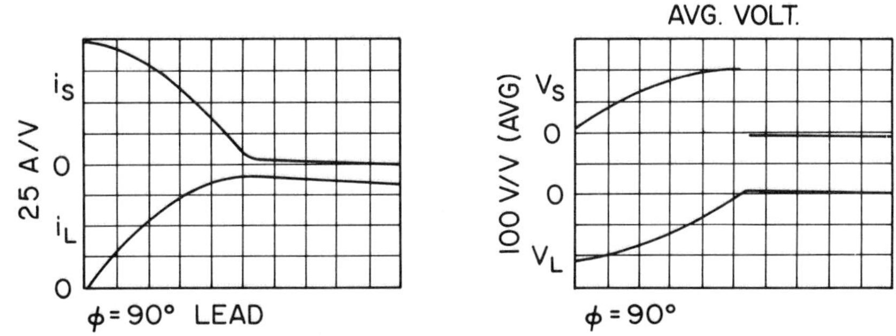

FIG. 8.4. Average ICB coil voltages and coil currents during a typical energy transfer cycle when $\phi = 90°$.

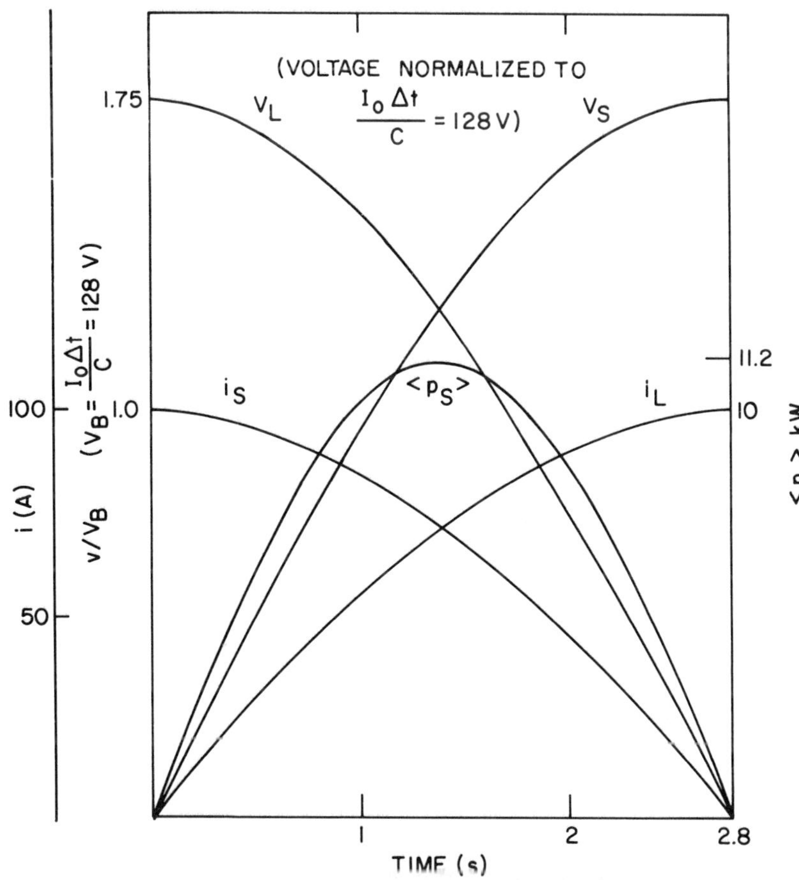

FIG. 8.5. Average ICB coil currents, voltages, and power vs. time.

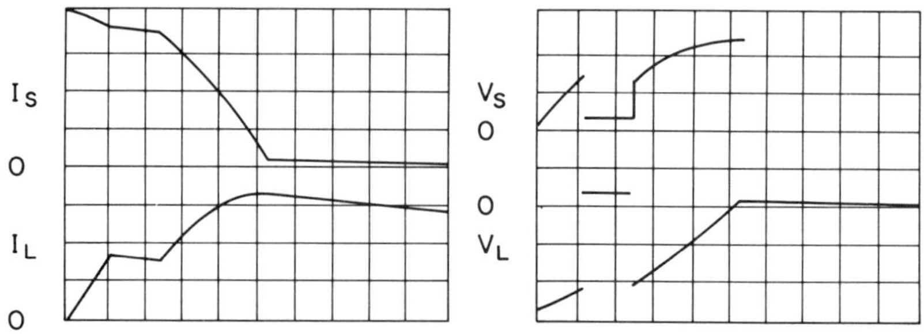

TOP STORAGE COIL CURRENT
BOT. LOAD COIL CURRENT
PHASE ANGLE 42°-0°-42° LEAD
1 DIV. = 25 amps 1 DIV. = 1.0 sec.

TOP AVE. STORAGE COIL VOLTAGE
BOT. AVE. STORAGE COIL VOLTAGE
PHASE ANGLE 42°-0°-42° LEAD
1 DIV. = 100 volts 1 DIV. = 1 sec.

FIG. 8.6. ICB system response to step function inputs to the microprocessor.

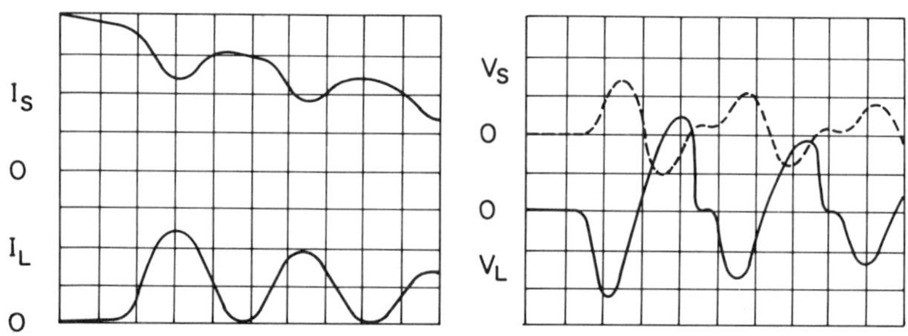

TOP CURRENT IN STORAGE COIL
BOT. CURRENT IN LOAD COIL
PHASE ANGLE ≈ ± 45°
FREQUENCY = 0.147 Hz
1.0 DIV.–25 A 1.0 DIV. = 2 sec.

TOP AVE. VOLT. ACROSS STOR. COIL
BOT. AVE. VOLT. ACROSS LOAD COIL
PHASE ANGLE ≈ ± 45°
FREQUENCY = 0.147 Hz
1.0 DIV. = 100 V 1.0 DIV. = 2 sec.

FIG. 8.7. ICB system response to a sine wave input to the microprocessor.

transfer. The current decay in the inductors during the $\phi = 0°$ intervals results from the losses. These losses primarily stem from the L/R decay in the holding-current shunt resistors and the converter losses which are almost at a constant rate.* No special attempts have been made, in the design of the prototype, to minimize converter losses. In the actual applications, loss minimization will be a primary concern.

*Some distortion of the scope CRT image is believed to be present because of the stray magnetic field from the storage and load coils nearby. Therefore, only rough signal measurements would be meaningful from these traces.

In Figure 8.7 a slow sinusoidal input was applied to produce sinusoidal variation of $\phi(t)$. Note the clipping effect in the storage coil current, I_S, when there is no more energy left in the load to be retrieved. The corresponding effect is also seen in I_L. Again, the pronounced decay of both currents is due to heavy system losses.

These experiments proved that the phase-shifting routine is a reliable service subroutine to be utilized in loop current control. Therefore, as soon as the $+90° \rightarrow -90°$ shift is completed, the bang-bang algorithm requires a $-90° \rightarrow +90°$ shift.

Figures 4.33 and 4.34 show a typical random reference signal and the corresponding system response, respectively. The optimal control of load current in bidirectional time was demonstrated by this simulation test.

The simulations of the ICB response established that bang-bang was a plausible control strategy. Experimental evaluation of the technique would determine the possible effects of the actual perturbations that were not accounted for in the computer model.

8.2.2 Closed-Loop Control Experiments

An overall diagram of the experimental closed-loop control setup is shown in Figure 8.8. The control algorithm block, in this figure, is detailed in Figure 7.1. The phase-shift algorithm is described in section 6.3.2. The ICB system block is shown in more detail in Figure 8.3.

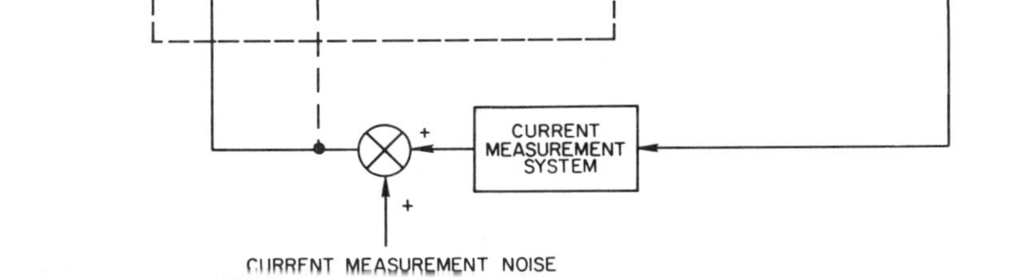

FIG. 8.8. Block diagram of the ICB closed-loop control system.

As has been mentioned, the microcomputer used in the experiments is an 8-bit system. It is shown in chapter 7 that an 8-bit digitizer with two's complement arithmetic will have a resolution of 0.781 A, in a 0–100 A range. Since this is the fundamental limit of the system resolution, common instrumentation practice is to design the peripheral instruments to be about ten times more accurate.

8.2.2.1 Bang-Bang Control Experiments. In a series of experiments the bang-bang control program was loaded in the microcomputer. A desired reference signal was supplied at the $i_R(t)$ channel of the A/D converter. The control program was scaled such that a 10 V amplitude of the reference signal was interpreted as 100 V. The results presented here are representative waveforms from these experiments.

Figure 8.9 shows a step 70 A reference and the corresponding load current response. Figure 8.10 shows the $i_L(t)$ dither for the same trace around the 70 A level in magnified time and amplitude scale. The expected bang-bang dither which was sketched in Figure 7.3 is quite evident. A typical peak to peak oscillation in $i_L(t)$ in this trace is about 900 mA, and the frequency of the dither is 22 Hz. This is consistent with the analytical and simulation predictions. The step response experiments established the validity of the bang-bang control algorithm experimentally. These experiments also showed the design accuracy of the control.

The control system response to various reference signals was demonstrated in a series of experiments in which rectangular, triangular, and sinusoidal signals were applied as $i_R(t)$. Figure 8.11 shows the response to a triangular reference of 10 A peak to peak superimposed on a 17 A dc level. The load current is seen to intercept the reference at $\phi = 90°$, under bang-bang command, and then proceed with the design dither about the reference.

If the triangular reference frequency is further increased, bang-bang control will spend a long time in $+90°$ or $-90°$ command, depending on the reference slope, Figure 8.12. If the reference frequency is too high, the system will no longer be able to follow it, as shown in Figure 8.13. It is obvious that losses will progressively reduce the frequency response of the system unless the losses are replenished, continuously, in the storage. Figure 8.14 shows the $i_L(t)$ response to a sinusoidal reference of 13.6 A peak to peak at a frequency of 0.28 Hz.

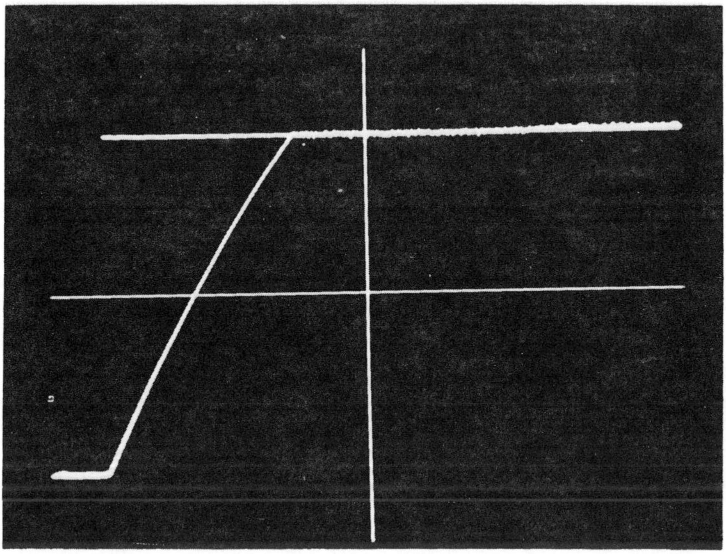

FIG. 8.9. FCB load current response to a 70-A stop reference, bang-bang control. The reference trace is superimposed on the response.

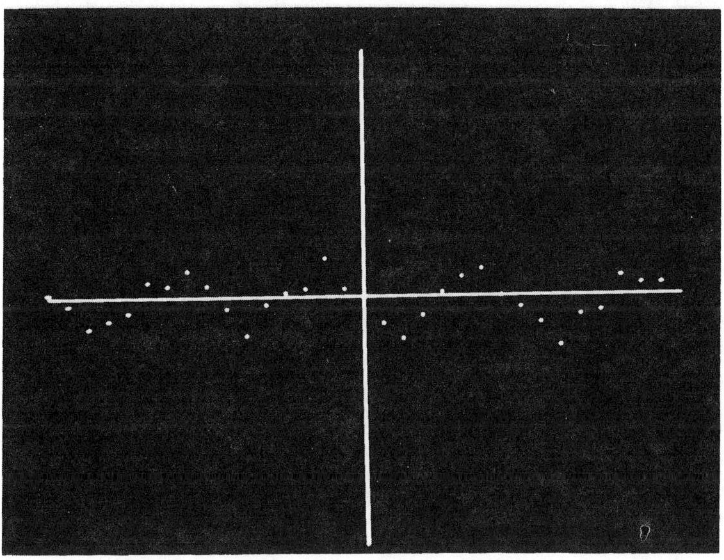

FIG. 8.10. Magnified view of i_L dither in Fig. 8.9.

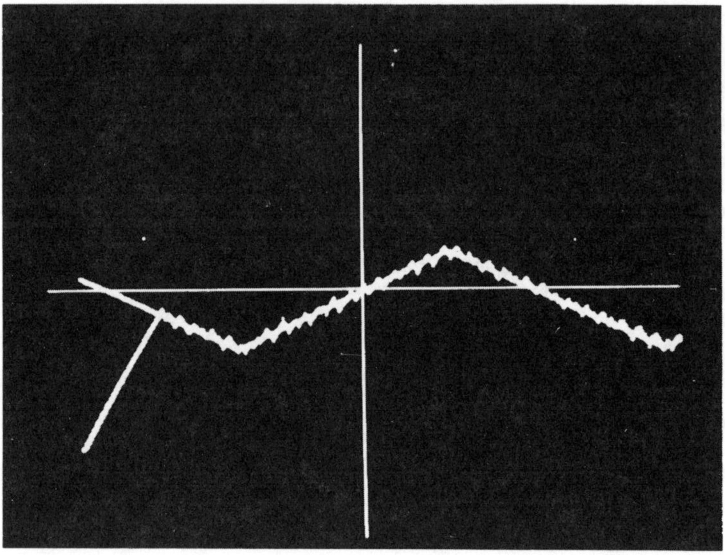

FIG. 8.11. Bang-bang response to a low-frequency triangular reference in the ICB. The reference and the i_L traces are superimposed.

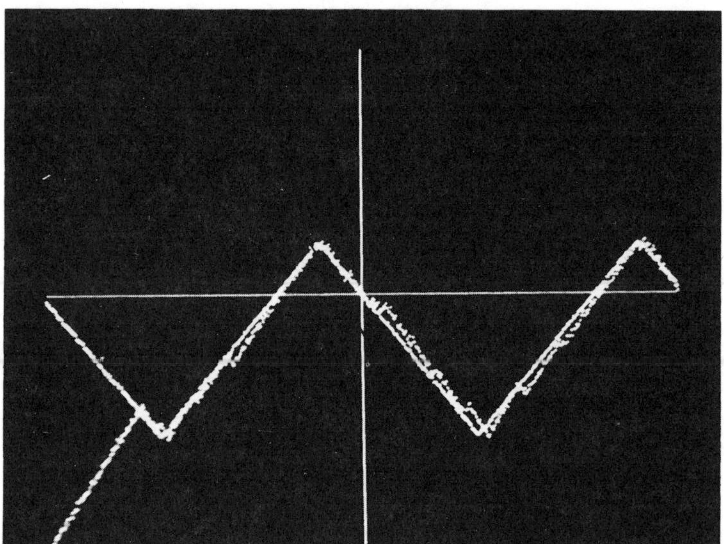

FIG. 8.12. Bang-bang response to a high-frequency periodic reference in the ICB. The reference and i_L traces are superimposed.

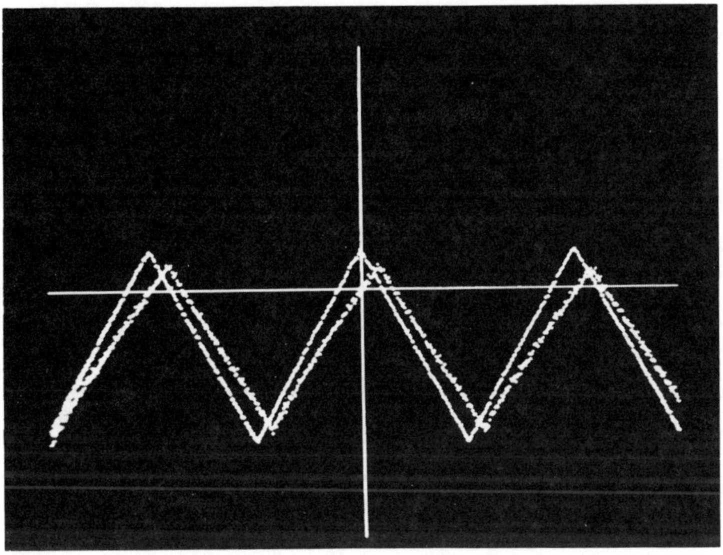

FIG. 8.13. Bang-bang response to a periodic reference with too high a frequency.

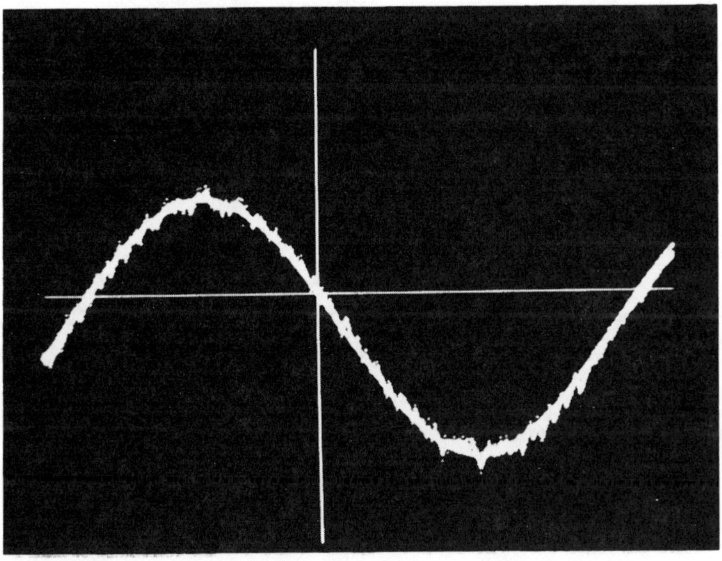

FIG. 8.14. Bang-bang response to a sinusoidal reference in the ICB. The reference and the i_L traces are superimposed.

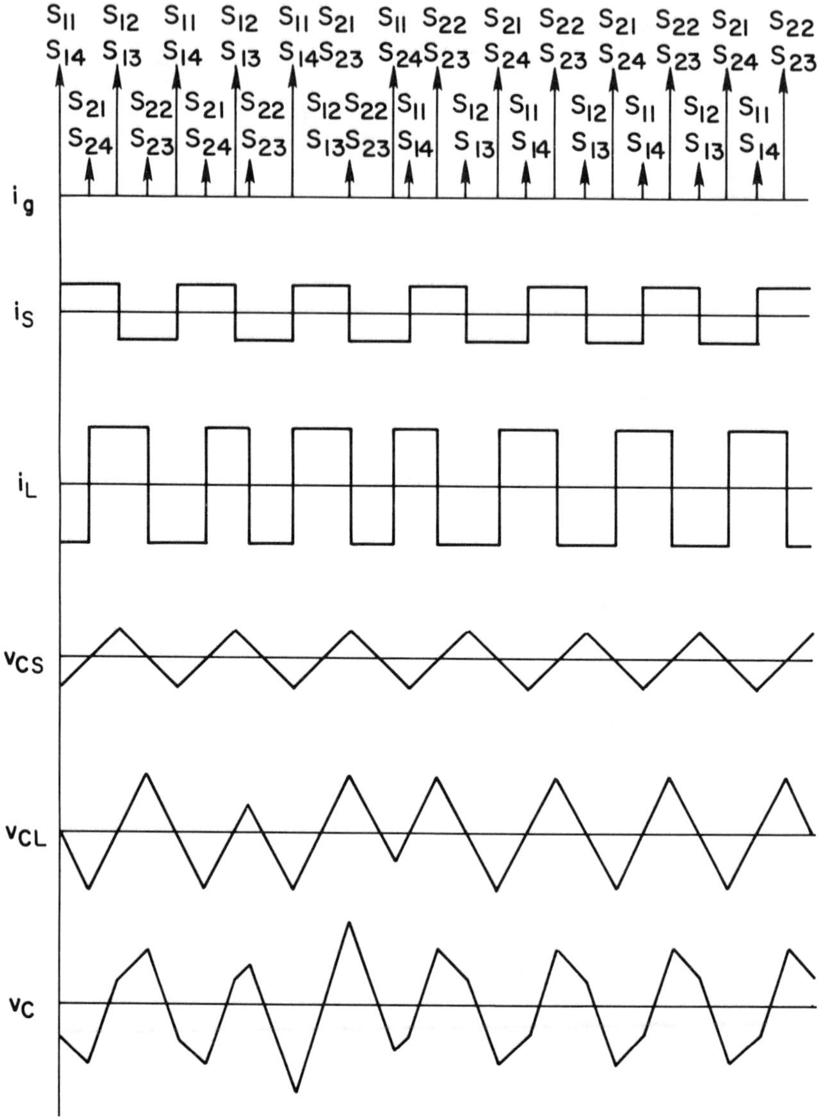

FIG. 8.15. Example of 180° phase shift in two 90° segments ($-90° \rightarrow 0° \rightarrow 90°$).

TABLE 8.4 The One-Phase Test Circuit

L_S	= 4 H
L_L	= 4 H
Converter frequency (f)	= 992 Hz
Phase capacitance (C)	= 2.5×10^{-4} F
SCR holding current (I_h)	= 0.8 A
Initial storage coil current (i_{S0})	= 100 A
Initial capacitor voltage (v_{C0})	= 220 V

If the reference frequency is increased too much, the load current again falls behind.

8.3 One-Phase ICB Experiments

A one-phase ICB experiment was set up to verify the operation and phase-shifting methods described in chapters 4 and 6. The one-phase ICB was adapted from the three-phase experimental ICB system which was already built at Argonne National Laboratory. The bang-bang control phase shift from 90° to −90° and vice versa was implemented in two three-step transient switching sequences for the one-phase ICB. Each three-step sequence accomplished a 90° shift. Figure 8.15 shows an example of such a two-segmented 180° phase shift. Some of the important test circuit values are listed in Table 8.4. A summary of the experimental results appears in the following section.

8.3.1 Constant-Phase Operation

An open-loop, constant-phase ($\phi = 60°$) test of the one-phase ICB was attempted to establish the viability of the circuit. The system operated successfully throughout a full energy transfer cycle from storage to load coil. Figure 8.16 shows the coil current waveforms through the transfer cycle. The (+) mark at the left shows the zero-time and zero-current point for the storage current, which is falling. The rising load current starts from the same time zero but is offset on the scope by about one division. It is observed that the currents follow a quarter-cycle of a sinusoid, which can be derived from equations (5.21) and (5.22), as evaluated with the circuit values. Figure 8.17 shows the capacitor voltage during the first few cycles of operation. This voltage was taken from one of the two series capacitors (C_1) in the adapted

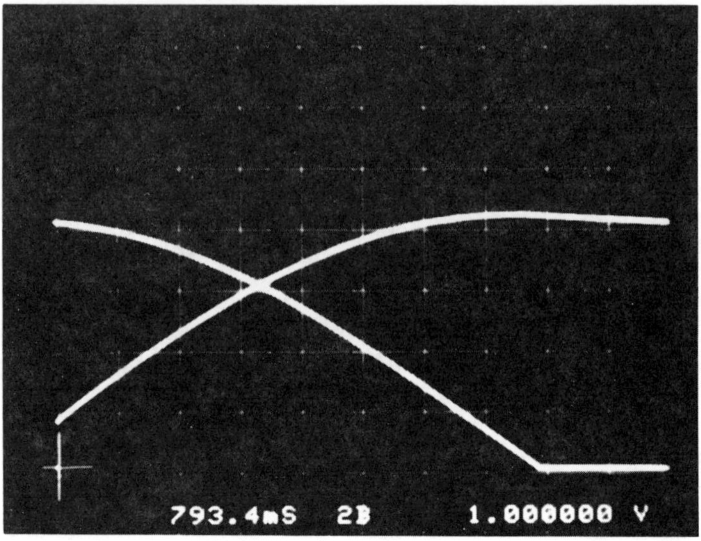

FIG. 8.16. Coil current waveforms through a full energy transfer at $\phi = 60°$. Scale: 25 A/div.

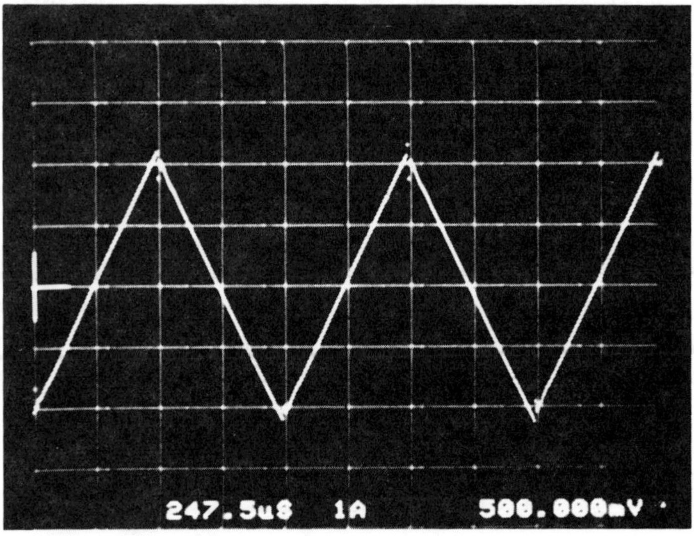

FIG. 8.17. Capacitor voltage during the first cycles of operation. Scale: 50 V/div.

one-phase ICB. The one-half precharge voltage of -110 V and storage-dominated waveform predicted in Figure 4.1 are evident. The small transients at the switching points show the effect of commutation loop currents which is present in every capacitor voltage trace. Figures 8.18, 8.19, and 8.20 show the capacitor and coil voltages at three different spots during the energy transfer. The traces from top to bottom, respectively, are v_S, v_C, and v_L. The (+) mark at the left in every case shows the zero voltage level of v_S. These waveforms are in close agreement with the prediction. For example, Figure 8.19 shows the waveforms near $i_S = i_L$, which is the same behavior as in Figure 4.3.

The constant-phase, full-energy transfer experiment was repeated at $\phi = 90°$. As was shown in the previous chapter, $\phi = 90°$ is on the threshold of commutation at the beginning of the cycle, $i_L/i_S = 0$. However, there is also no current in L_L to be commutated. As the net v_L voltage, shown in Figure 4.2, establishes some i_L current, the threshold phase goes beyond 90° and $\phi = 90°$ commutation progressively improves. These arguments were supported by the success of energy transfer at a constant $\phi = 90°$. Figure 8.21 shows an example

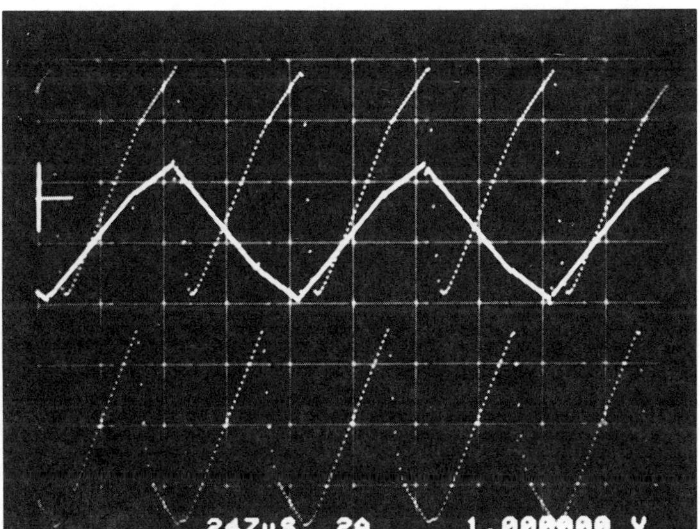

FIG. 8.18. Capacitor and inductor voltage waveforms during t ≃ 1s at $\phi = 60°$. Top, v_S, middle, v_C; bottom, v_L.

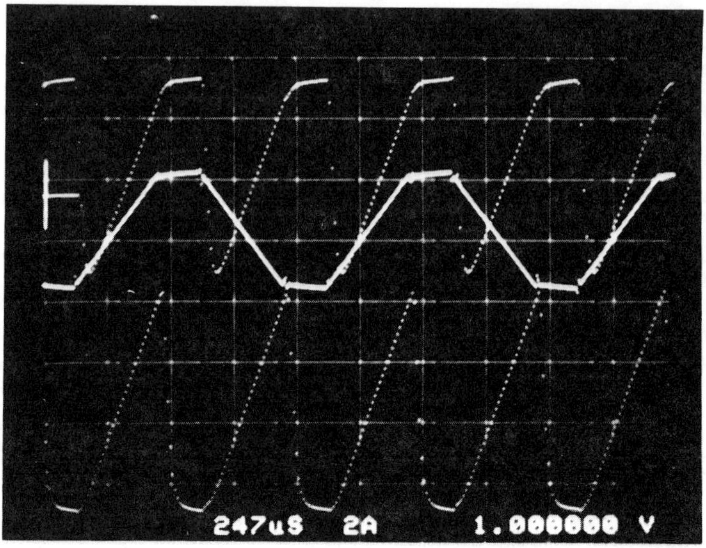

FIG. 8.19. Capacitor and inductor voltage waveforms during t $\simeq$ 2.5s at $\phi = 60°$. Top, v_S; middle, v_c; bottom, v_L.

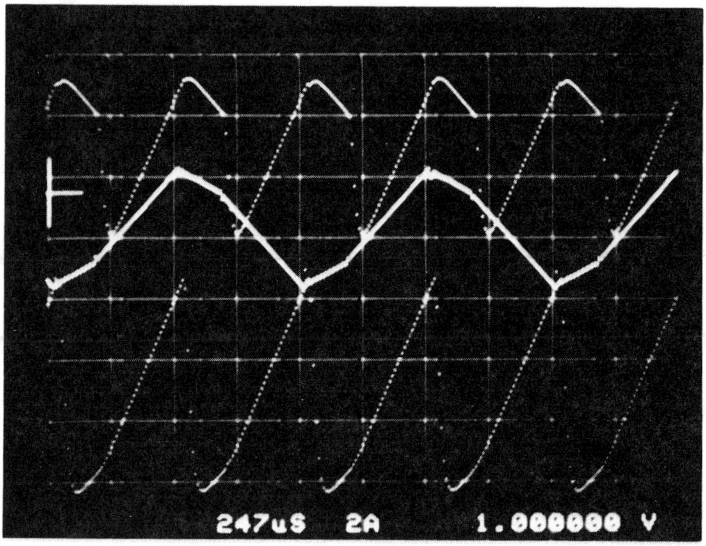

FIG. 8.20. Capacitor and inductor voltage waveforms during t $\simeq$ 4s at $\phi = 60°$. Top, v_S; middle, v_c; bottom, v_L.

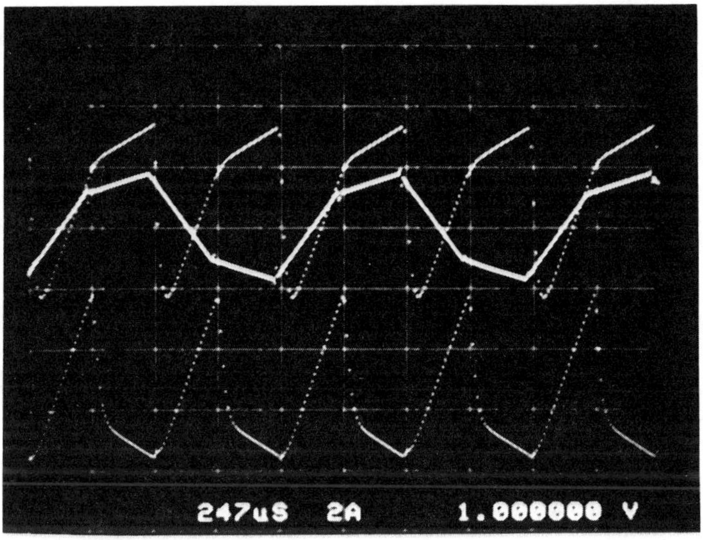

FIG. 8.21. Capacitor and inductor voltage waveforms during t ≃ 2s at $\phi = 90°$. Top, v_S; middle, v_c; bottom, v_L.

of the inductor and capacitor voltages. The $\phi = 90°$ phase relationship in this oscillogram can be verified by the fact that the storage and coil voltage switchings are occurring at equal intervals.

The three-step phase-shifting algorithm was tested in a closed-loop bang-bang control experiment. The 180° phase shifts were done in two three-step 90° shift segments, as was shown in Figure 8.15. Figure 8.22 shows an example of the capacitor voltage during a $\phi = -90°$ to $\phi = 90°$ phase shift. For better understanding of how this waveform is produced, Figure 8.15 should be studied. The straight line connecting the highest two capacitor voltage peaks shows the switching interval in which the two converters are in phase ($\phi = 0$). Figure 8.23 shows the inductor voltages during the same phase shift as that in Figure 8.22. The change of average load voltage from positive to negative can be seen as the load voltage oscillations (top trace) go through the shifting transient steps. A similar change of net storage voltage from negative to positive can be seen in the bottom trace. The switching interval with $\phi = 0°$ is again seen in both traces as a straight line. The zero voltage axis of each trace runs through the midpoint of this line.

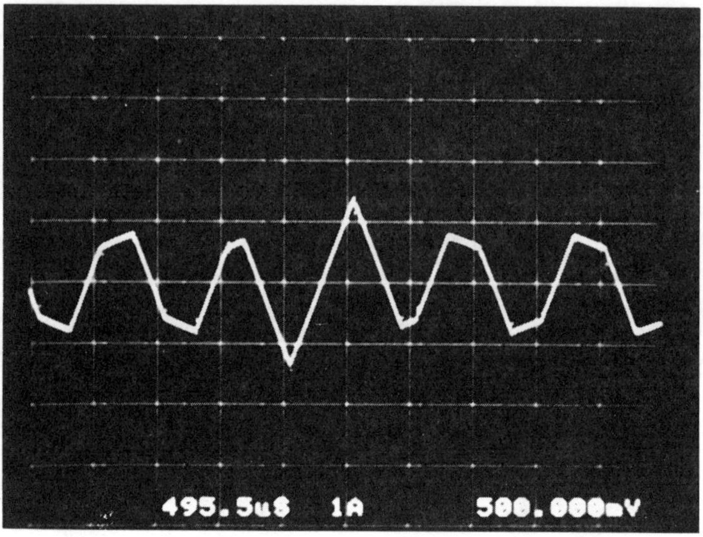

FIG. 8.22. Capacitor voltage during a $\phi = 90°$ to $-90°$ phase shift.

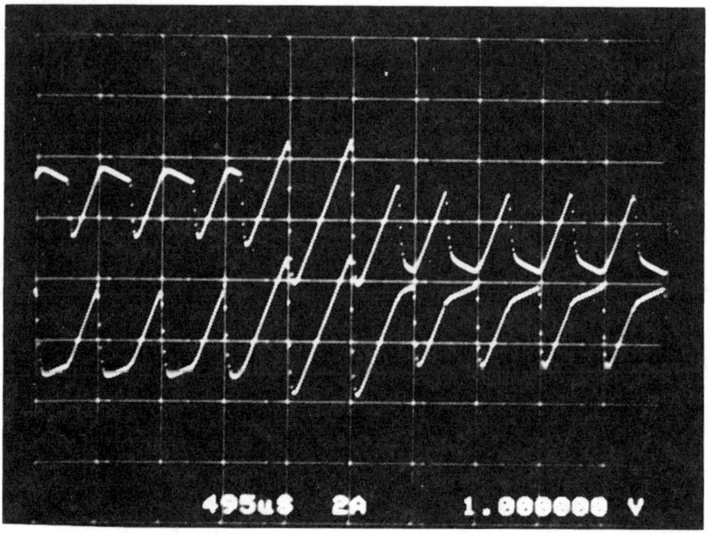

FIG. 8.23. Coil voltages during a $\phi = 90°$ to $-90°$ phase shift.

The experimental evaluation of the one-phase ICB established the practicality of this system as a controllable two-quadrant power supply for superconducting magnets.

References

Ehsani, M., and R. L. Kustom. *A New One-Phase Dual Converter for Superconducting Inductive Energy Storage and Transfer Applications: The One-Phase Inductor-Converter Bridge*. Argonne National Laboratory Technical Report ANL/FPP/TM-182. Argonne, Ill.: Argonne National Laboratory, March 1984.

———, R. L. Kustom, and R. E. Fuja. "Microcomputer Control of a Current Source DC-DC Converter." *IEEE Transactions on Industry Applications*, IA-19, no. 5 (September/October 1983).

Fuja, R. E.; R. L. Kustom; and M. Ehsani. "Three-Phase Energy Transfer Circuit with Superconducting Energy Storage Coils." *IEEE Transactions on Industry Applications*, IA-16, no. 3 (May/June 1980).

Stein, P. K. *Measurement Engineering*, vol. I, *Basic Principles*. Phoenix, Ariz.: Stein Engineering Services, 1964.

CHAPTER 9

Converters for Diurnal Superconductive Magnetic Energy Storage

The converter circuits discussed thus far in the book have been for energy transfer between two superconductive magnets. In these systems one coil was for the production of a desired magnetic field, while the other coil served as the energy storage buffer. The primary application of these systems is in experimental research in which high magnetic fields are required.

In this chapter we will discuss another application of converter circuits which interface superconductive magnets with the conventional electric power utility system. The high energy density of the superconductive magnet and the quick response of the solid-state-switching power converter are utilized in this application to produce a flexible bulk energy storage and transfer for the power system.

9.1 Diurnal Energy Storage

Power demand on an electric utility varies substantially over a twenty-four-hour cycle. Some typical cycles are shown in Figure 9.1. Utilities handle this variation with different generating units, starting with base-load units that operate at nearly constant energy output over the twenty-four-hour period. These units are the most efficient and operate at the lowest fuel cost. As the power demand increases, intermediate coal or oil units are brought on line. The fuel costs are usually several times more than those of a base-load unit. Finally, for short-term peaks, gas or diesel units are brought on line. The fuel costs for these units are over an order of magnitude greater than those of base-load units.

Pumped hydroelectric systems have been built to store energy at night and discharge it during the day to make more efficient use of baseload units. However, pumped hydroelectric power is geographically limited and is only about 60–75% efficient. Boom and Peterson have suggested the use of superconductive magnetic energy storage

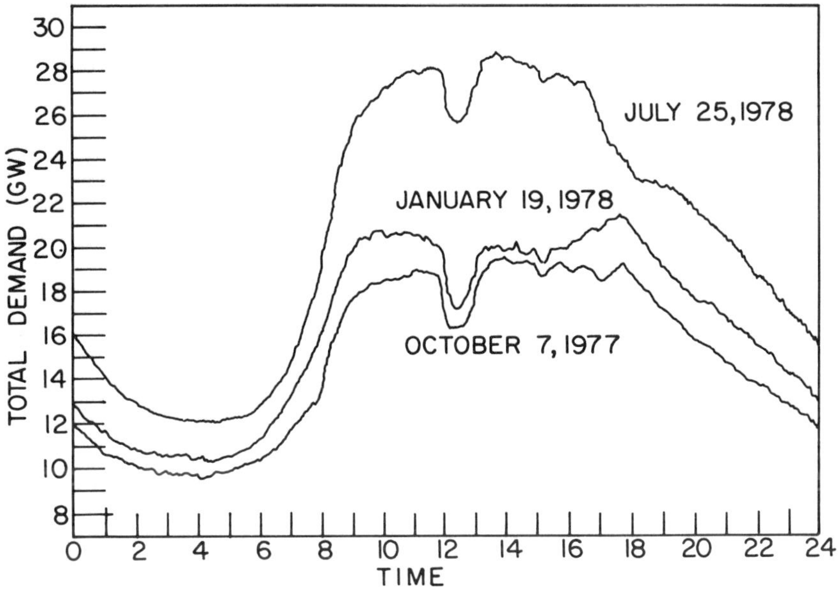

FIG. 9.1. Typical power demand curves on an electric utility over a 24-hour period.

(SMES) for this application. SMES can be more than 95% efficient and can be almost universally sited. Research into achieving an economic design for such a system is in progress.

Energy in an SMES unit is inductively stored; i.e.,

$$w_d(t) = \frac{1}{2} L_d I^2(t)$$

where $w_d(t)$ is the energy, L_d is the coil inductance, and $I(t)$ is the current.

Energy is added to the coil during the nighttime, when power demand is low, and is extracted during the daytime, when power demand is high. The current in the coil slowly cycles from the minimum current, I_{min}, to the maximum current, I_{max}, and back to I_{min} over the twenty-four-hour cycle.

The most economic sizes of SMES appear to be greater than 1,000 mwh. The coils that are proposed for this application have low aspect ratios (i.e., low height to radius), single-layer construction, and high operating currents. A typical unit would have a radius of several hundred meters, a height of a few tens of meters, an operating current of several hundred thousand amperes, and an inductance of tens of henries. The proposed conductor is a niobium titanium superconductor with a composite of pure aluminum for stabilization and aluminum alloy for structural strength.

9.2 Interaction with ac Utilities

The proposed interface between the SMES coil and the ac utility is a Graetz bridge. The circuit diagram for a six-pulse bridge is shown in Figure 9.2. The general behavior of this circuit, connected to an SMES coil, is identical to that of a converter connected to a dc transmission system with a large smoothing reactor. A gate pulse is delivered to one of the SCRs every 60° of cycle. The conducting sequence follows the same pattern that was previously described for the three-phase ICB circuit. Commutation of current from one SCR to another can take place at any time when the line-to-line voltage is positive on the anode of the SCR that is to start conduction. Therefore, the successful choice of firing times can be selected over a half-cycle.

The utility, by virtue of its large installed generating capacity, acts

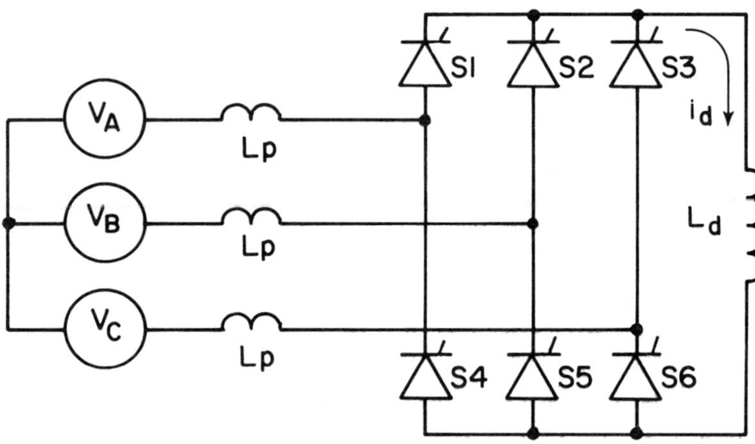

FIG. 9.2. Circuit diagram for a six-pulse Graetz bridge connected to an SMES coil.

as a constant voltage source. The superconducting coil has sufficiently large stored energy that changes in the operating current from minimum to maximum value take hours. Thus, the SMES coil acts as a current source the value of which changes very slowly over time and is virtually constant on the time scale of ac cycles. Unless a tap-changing transformer is used to interface the converter with the utility, or a current bypassing circuit is used across the coil, the volt-amperes seen by the ac system are constant at any given time in the diurnal cycle, irrespective of the power being delivered to or taken from the SMES coil.

9.3 Graetz Bridge Behavior with an SMES Coil

In a six-pulse Graetz bridge, gate pulses are delivered at regular intervals every one-sixth of an ac cycle. The conducting pairs in Figure 9.2 rotate from S1S5 to S1S6 to S2S6 to S2S4 to S3S5, etc. Consider the three voltages V_A, V_B, and V_C.

$$V_A = V_m \sin \omega t,$$

$$V_B = V_m \sin(\omega t - 120°),$$

and

$$V_C = V_m \sin(\omega t - 240°). \tag{9.1}$$

The commutation of current from one SCR to another requires the right voltage polarity in the commutating loop. For instance, for current to commutate from S3 to S1, the voltage in the V_A, S1, S3, and V_C loop must provide a positive polarity on the anode of S1 and a negative polarity on the anode of S3.

The voltage across S1, V_{S1}, is

$$V_{S1} = V_A - V_C,$$

$$V_{S1} = V_m \sin \omega t - V_m \sin(\omega t - 240°),$$

$$V_{S1} = \sqrt{3} V_m \sin(\omega t - 30°). \tag{9.2}$$

The line-to-line voltage, V_{AC}, is positive for 180° between ωt equal to 30° and 210°. Gating pulses during that part of the cycle will drive the current from phase C to phase A and from S3 to S1. If ωt is replaced by $\alpha + 30°$, then

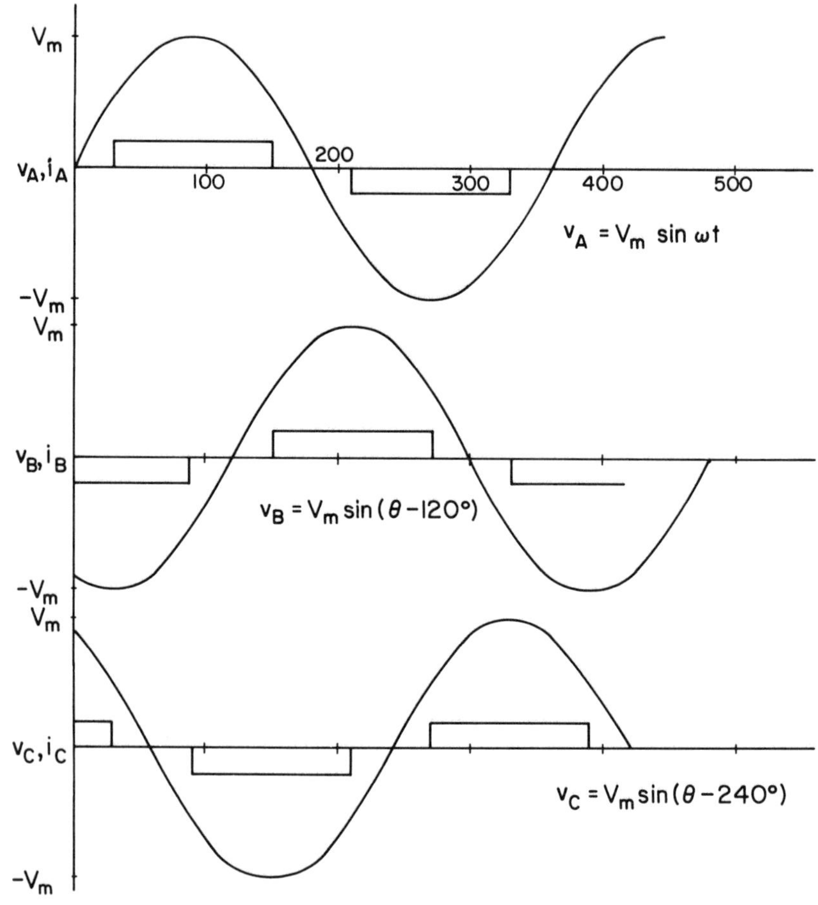

FIG. 9.3. Three phase voltages and currents for a Graetz bridge connected to an SMES coil.

$$V_{S1} = \sqrt{3} V_m \sin \alpha, \tag{9.3}$$

where α is defined as the delay angle. Commutation can occur for any value of α between 0° and 180°. Figure 9.3 shows the three phase voltages and currents for a Graetz bridge connected to an SMES coil when the phase reactances, L_p, are effectively zero.

It should be noted that the continuity of coil current, I_d, must be and is always conserved in one switching cycle.

Ignoring phase reactance for the moment, the voltage on the SMES coil, V_d, is a series of piecewise line-to-line voltage segments given by

$$V_d = \sqrt{3}V_m \cos(\theta + \alpha), \qquad -\pi/6 \geqslant \theta \leqslant \pi/6. \tag{9.4}$$

The average coil voltage is

$$\langle V_d \rangle = \frac{3}{\pi}\int_{-\pi/6}^{\pi/6} \sqrt{3}V_m \cos(\theta + \alpha)d\theta$$

$$\langle V_d \rangle = \frac{3\sqrt{3}}{\pi} V_m \cos \alpha. \tag{9.5}$$

The SMES coil current is effectively constant over a cycle; therefore, the average power into or out of the coil is the product of the average coil voltage and the current, i.e.,

$$P_d = \langle V_d \rangle I_d,$$

$$P_d = \frac{3\sqrt{3}}{\pi} V_m I_d \cos \alpha. \tag{9.6}$$

If α is between 0 and $\pi/2$, average power is delivered to the coil, the coil current slowly increases, and so does the stored energy. If α is between $\pi/2$ and π, average power is taken from the coil, the coil current slowly decreases, and so does the stored energy.

Current in the individual phases is shown in Figure 9.3. Fourier analysis of the current gives for the fundamental component:

$$I_1 = \frac{2\sqrt{3}}{\pi} I_d \sin(\omega t - \alpha). \tag{9.7}$$

The average power flowing on the ac side is three times the average power per phase, i.e.,

$$\langle P_{AC} \rangle = 3\left\{\frac{1}{2\pi}\int_0^{2\pi} (V_m \sin\theta)\left(\frac{2\sqrt{3}}{\pi}\right) I_d(\sin\theta - \alpha)d\theta\right\}$$

$$\langle P_{AC} \rangle = \frac{3\sqrt{3}}{\pi} V_m I_d \cos \alpha . \tag{9.8}$$

Equations (9.8) and (9.6) are identical, since the process of transferring energy between the utility and the coil has not included any circuit losses.

The phase reactance prevents immediate commutation between SCRs, and as a result for a period of time three SCRs are conducting simultaneously. This condition is treated in detail in textbooks on dc transmission. A brief summary is presented here.

Figure 9.4 indicates the condition in which the coil current at the start of commutation is flowing in phases A and C and the current is to be commutated from A to B. A loop current, i_L, starts to flow when S2 is gated on. The phase inductances in A and B limit the rate of rise of the loop current. When i_L reaches the value of I_d, then all the coil current is in phase B and none in phase A. Commutation is complete. During this period, both S1 and S2 are conducting. At the end of the period, S1 is no longer conducting because reverse current cannot flow in S1, and S2 is the only SCR conducting. The period when both are conducting is known as the overlap period, and the angular duration of overlap is known as the overlap angle. During the overlap period,

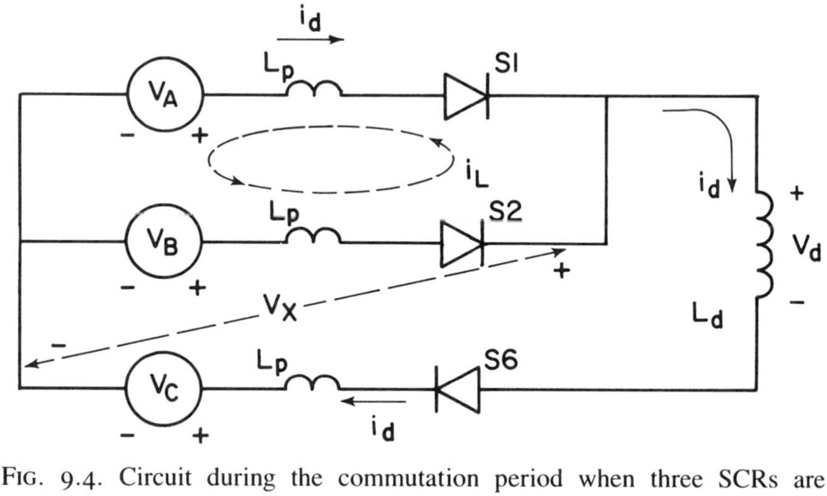

FIG. 9.4. Circuit during the commutation period when three SCRs are conducting.

the voltage on the SMES coil is equal to $V_x - V_c$. Equation (9.4) is not correct during the overlap period but is correct otherwise.

The correction for the overlap period will be treated as a perturbation to equation (9.5). During the overlap period, the coil voltage is given by

$$V_d = V_x - V_c \tag{9.9}$$

and

$$V_x = V_B - L_p\frac{di}{dt} = V_A + L_p\frac{di_L}{dt}. \tag{9.10}$$

$$V_B - 2L_p\frac{di_L}{dt} - V_A = 0 \tag{9.11}$$

The voltage in the commutation loop at the instant commutation starts is given by equation (9.3), $\sqrt{(3)}V_m \sin\alpha$, and at the end of overlap is $\sqrt{(3)}V_m \sin(\alpha + \mu)$ where μ is the overlap angle. Thus, in equation (9.11)

$$2L_p\frac{di_L}{dt} = \sqrt{3}V_m \sin\omega t$$

and

$$\int di_L = \frac{1}{2L_p}\int_\alpha^{\alpha+\mu}\sqrt{3}V_m \sin\omega t\left(\frac{d\omega t}{w}\right),$$

$$i_L(\alpha + \mu) - i_L(\alpha) = \frac{\sqrt{3}V_m}{2\omega L_p}[\cos\alpha - \cos(\alpha + \mu)]. \tag{9.12}$$

Defining

$$I_L = \frac{\sqrt{3}V_m}{2\omega L_p}, \tag{9.13}$$

and since $i_L = 0$ at the start of commutation, i.e., $i_L(\alpha) = 0$, and $i_L = I_d$ at the end of commutation, i.e., $i_L(\alpha + \mu) = I_d$, then

$$\frac{I_d}{I_L} = \cos\alpha - \cos(\alpha + \mu). \tag{9.14}$$

The length of the overlap period is defined by equation (9.14).

The average coil voltage given by equation (9.5) must be corrected to account for the overlap period. The correction will be denoted by ΔV, i.e.,

$$\langle V_d \rangle = E_{dm} - \Phi \Delta V,$$

where

$$E_{dm} = \frac{3\sqrt{3}}{\pi} V_m \cos\alpha. \tag{9.15}$$

The value of V_x is found from equations (9.10) and (9.11), i.e.,

$$V_x = 0.5(V_A + V_B). \tag{9.16}$$

During the overlap period, the coil voltage is

$$V_d = 0.5(V_A + V_B) - V_c, \tag{9.17}$$

where normally the coil voltage is

$$V_d = V_A - V_c. \tag{9.18}$$

The difference in voltage between equations (9.18) and (9.17) is $0.5(V_A - V_B)$, which from equations (9.11) and (9.12) is

$$0.5(V_A - V_B) = \frac{\sqrt{3}}{2} V_m \sin\theta, \qquad \alpha \leqslant \theta \leqslant \alpha + \mu. \tag{9.19}$$

This perturbation must be added to the integration performed in equation (9.5). Essentially

$$\Delta V = \frac{3}{\pi} \int_{\alpha}^{\alpha+\mu} \frac{\sqrt{3}}{2} V_m \sin\theta \, d\theta$$

$$\Delta V = \frac{3\sqrt{3}}{2\pi} V_m [\cos\alpha - \cos(\alpha + \mu)]. \tag{9.20}$$

The term in the brackets is the same as in equation (9.14). Therefore, equation (9.20) is

$$\Delta V = \frac{3\sqrt{3}}{2\pi} V_m \frac{I_d}{I_L},$$

$$\Delta V = \frac{3I_d}{\pi}(\omega L_p). \tag{9.21}$$

In dc transmission analysis, the following unit definitions are frequently applied:

$$E_{dm} = \frac{3\sqrt{3}}{\pi} V_m,$$

$$Z_b = \frac{\pi E_{dm}}{6I_{dm}},$$

and

$$I_{dm} = \text{maximum current rating of the coil,}$$

and

$$E_d = \langle V_d \rangle. \tag{9.22}$$

Then the coil voltage is

$$\langle V_d \rangle = \frac{3\sqrt{3}}{\pi} V_m \cos\alpha - \frac{3I_d}{\pi}(\omega L_p)$$

$$E_d = E_{dm}\cos\alpha - \left(\frac{\omega L_p}{2Z_B}\right)\left(\frac{I_d}{I_{dn}}\right) E_{dm}. \tag{9.23}$$

Defining the unit quantities $I_d' = I_d/I_{dn}$, $X_p' = \omega L_p/Z_B$,

$$E_d = E_{dm}\left(\cos\alpha - \frac{X_p' I_d'}{2}\right). \tag{9.24}$$

9.3.1 Voltages, Currents, Real and Reactive Power

The size of the SMES coil is determined by the size of the utility and its daily power fluctuations. Typical units would store several thousand megawatt-hours to productively reduce or remove the diurnal fluctuations. In reference to Figure 9.1, the most likely operating scenario would be to charge during the night starting from nearly zero power, peaking sometime in the middle of the night, and returning to low or near zero power towards morning. The daytime output would start near zero power, peak rapidly at a somewhat lower power level than the nighttime peak, and staying near that level for a longer time period than the nighttime charge. Superimposed on the daytime period would be short-term peaks to provide short-term surges of about an hour or less. An idealized power curve could look something like Figure 9.5.

The voltage and current are determined by the coil design. Obviously, the $\oint P_d dt$, where P_d is the coil power, over a twenty-four-hour cycle is zero, and

$$\int_{\text{night}} P_d dt = -\int_{\text{day}} P_d dt = \frac{1}{2} L_d I_d^2(\max) - \frac{1}{2} L_d I_d^2(\min), \quad (9.25)$$

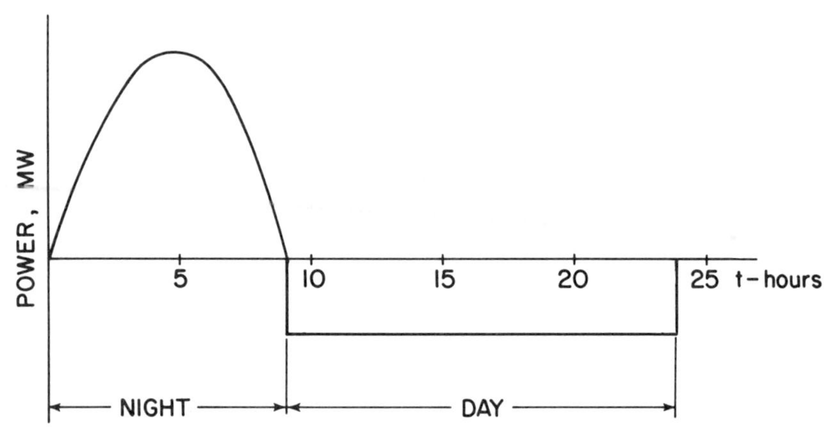

FIG. 9.5. Idealized diurnal power cycle for an SMES unit.

where $I_d(\max) = I_{dn}$, the maximum design current of the coil, and $I_d(\min) =$ the minimum operating current of the coil.

The required line voltage is determined from equations (9.22) and (9.24) and the peak power and the coil current at the peak power. The coil current at peak power might or might not be equal to the peak coil current; it is not, for instance, in the idealized power curve shown in Figure 9.5.

At first assume that $X_p' = 0$. Then the maximum dc coil voltage, $E_d(\max)$, is equal to E_{dm} for α equal to zero. The maximum coil current is I_{dn}. The line-to-neutral voltage, V_m, and line current, I_m, are

$$V_m = \frac{\pi}{3\sqrt{3}} E_{dm}, \tag{9.26}$$

and

$$I_m = \frac{2\sqrt{3}}{\pi} I_{dn}. \tag{9.27}$$

The total three-phase rms volt-amperes is

$$VA = \frac{3}{2} V_m I_m,$$

and

$$VA = E_{dm} I_{dn}. \tag{9.28}$$

The real and reactive power on the ac side if α is varied is

$$P = E_{dn} I_{dn} \cos \alpha,$$

and

$$Q = E_{dm} I_{dn} \sin \alpha. \tag{9.29}$$

The values of real and reactive power are the locus of points on a circle with the radius equal to the product of the voltage, E_{dm}, and the coil current, I_d. Therefore,

$$(VA)^2 = E_{dm}^2 I_d^2 = p^2 + Q^2. \tag{9.30}$$

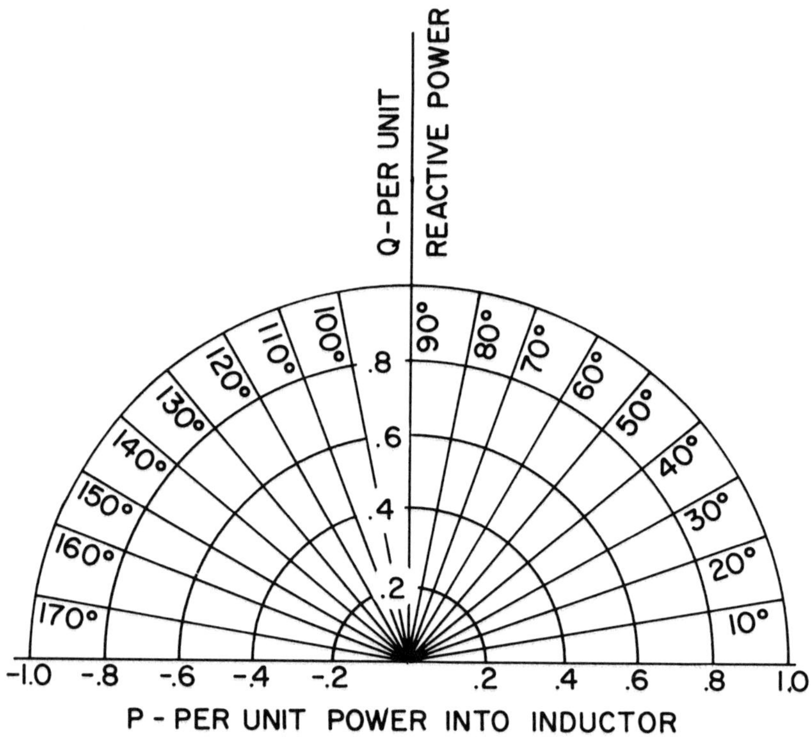

FIG. 9.6. Normalized circle diagram for P and Q when $X'_p = 0$.

If the voltage is normalized to E_{dm} and the current to I_{dn}, then the maximum radius of the cycle in equation (9.30) is unity when I_d is at the full design current, I_{dn}. The loci of points for P and Q at other values of I_d are circles with radii equal to the fraction I_d/I_{dn}. Normalized circle diagrams when $X_p' = 0$ are shown in Figure 9.6.

If X_p' is not equal to zero, then E_{dm} has to be increased according to equation (9.24) to match the peak power requirements. Another effect of a finite X_p' is that the current in the ac line is phase-shifted with respect to the voltage approximately by the sum of the overlap angle, μ, and the delay angle, α. If the peak power is defined as the per-unit base, then the value of volt-amperes in that base, $(VA)_{pu}$, is greater than unity. That is,

$$(VA)_{pu} = \left(\frac{E_{dm}}{E_{dn}}\right)\left(\frac{I_{dn}}{I_d(P = \max)}\right), \tag{9.31}$$

where E_{dn} is the maximum value of E_d given by equation (9.24) for the value of $I_d[I_d(P = \max]/I_{dn})$ at the maximum power value and the minimum value of α (nearly zero) at which the bridge will be operated at full power.

The idealized power cycle given in Figure 9.5 is used as an example for the diurnal variation of coil current, coil voltage, and real and reactive power in per-unit quantities of E_{dn} and I_{dn}. The results are shown in Figure 9.7.

Several conclusions are evident. If the power cycle follows availability of power faithfully, then peak power and peak current do not coincide. The value of reactive power is extremely large when the power level is low. The former presents no problem, and the latter

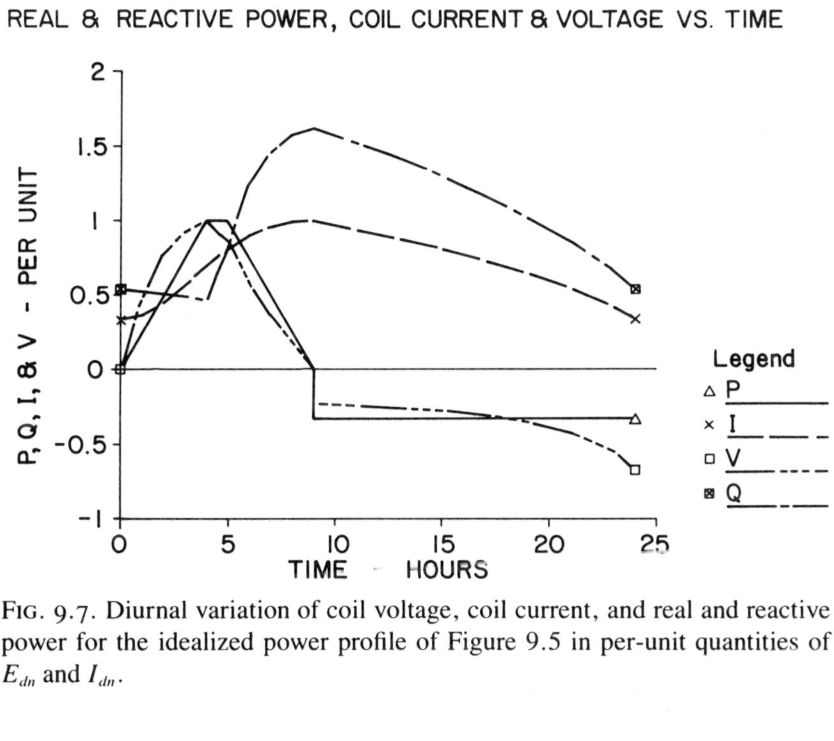

FIG. 9.7. Diurnal variation of coil voltage, coil current, and real and reactive power for the idealized power profile of Figure 9.5 in per-unit quantities of E_{dn} and I_{dn}.

requires correction. The subject of reactive power control is addressed in section 9.5.

9.3.2 Component Ratings

Under steady-state conditions, the coil current flows through any given SCR for one-third of a cycle. The rms rating of each switch element in the bridge is:

$$I_R = \left[\frac{1}{2\pi}\int_0^{2\pi/3} I_{dn}^2\, d\theta\right]^{\frac{1}{2}},$$

$$I_R = \frac{I_{dn}}{\sqrt{3}}. \tag{9.32}$$

Any single switch element in the bridge is connected across the coil terminals at one point in the cycle. Since the delay angle is operated over virtually the full 180° in SMES operation, the forward and reverse voltages across the switch elements can be equal to the full line-to-line voltage, i.e., $\sqrt{(3)}V_m$.

Since an SMES system is directly connected to the utility system, the bridge must be designed not only for steady-state operation, but also for fault conditions. In practice, the $\int I^2 dt$ rating of the SCRs must match the short-circuit rating of the transmission line, and the voltage rating must be chosen to allow for the overvoltage surges that can exist with the design of the transient voltage protection devices on the transmission line.

9.4 Current Sharing Between Parallel Bridge Units

The technique for current sharing described in chapter 1 can be applied for the SMES circuits; however, because the coil current is envisioned to be so large, it is desirable to divide the bridge into many parallel units. The control of current sharing between parallel units can be accomplished through delay control, as is done with multiterminal dc transmission systems.

The possible sources of imbalance in parallel systems can include slightly different terminal voltages because of transformer variations, differences in SCR voltage drops, and different lead voltage drops and phase shifts in the ac lines between parallel units. Figure 9.8 depicts a

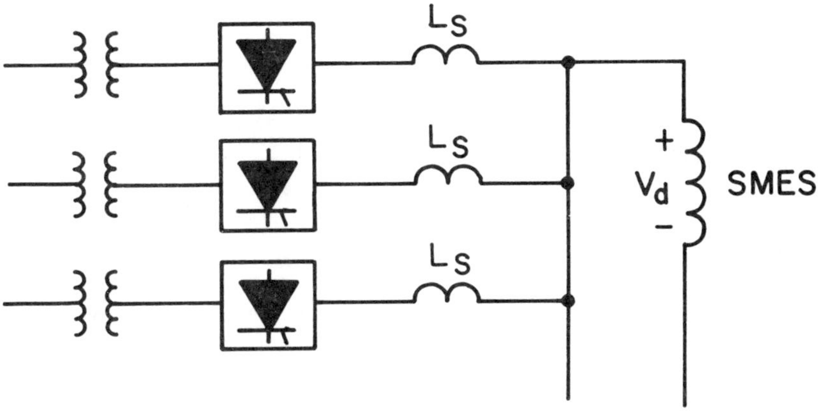

FIG. 9.8. Parallel connection of multiple converter units with smoothing reactors.

large number of parallel-connected converter units, each with a series-smoothing reactor. Ideally, the current in each unit is Idn/n. The imbalances in individual converter voltages are to be corrected by delay angle control, and the smoothing reactor is to prevent short-term current spikes.

Depicted in Figure 9.9 is the instantaneous SMES coil voltage for an arbitrary value of delay angle, α. This voltage is

$$V_d = \sqrt{3}V_m \cos(\theta + \alpha),$$

where

$$-\pi/6 \leqslant \theta \leqslant \pi/6. \tag{9.33}$$

As indicated in Figure 9.9, the voltage just before switching is different from the voltage just after switching. If one of the bridge units is established as the reference unit, then its delay angle can be chosen to achieve the right SMES coil voltage, and the delay angles on other bridges can be adjusted later or earlier to keep current evenly divided among the bridges.

A voltage difference, ΔV, before and after switching is

$$\Delta V = \sqrt{3}V_m[\cos(\pi/6 + \alpha) - \cos(\pi/6 - \alpha)],$$

$$\Delta V = -3V_m \sin\alpha. \tag{9.34}$$

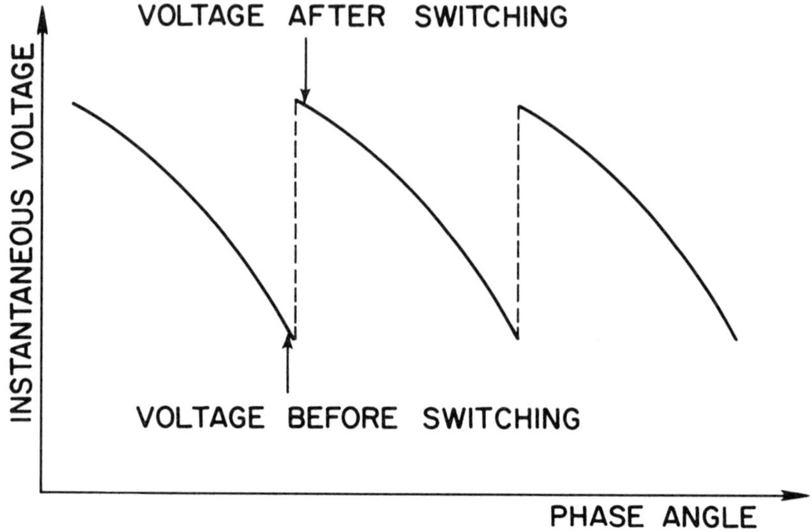

FIG. 9.9. Instantaneous SMES coil voltage for an arbitrary value of α.

Consider the case for two bridges; if the delay angle on one bridge is delayed more than the other by an amount, $\Delta\alpha$, then a voltage spike is generated between the two bridges in the amount of equation (9.34). The voltage spike will last for a period of time, $\Delta\alpha/\omega$. During the voltage spike, a loop current between the two bridges will be started that increases by the amount

$$\Delta i_L = \frac{\Delta V}{2L_s}\left(\frac{\Delta\alpha}{\omega}\right),$$

$$\Delta i_L = \frac{-3V_m}{2\omega L_s}\Delta\alpha \sin\alpha. \tag{9.35}$$

The current in one bridge will be increased by the amount, Δi_L, and the current in the other bridge will decrease by that amount. This process occurs six times every ac cycle. At the same time, the voltages in two bridges are assumed not to be equal; otherwise the two currents would be identical and delay angles would be the same ($\Delta\alpha = 0$). Therefore, over the whole cycle an average voltage error exists in the amount $\langle \Delta V \rangle$ because of the inherent imbalance. This imbalance would lead to a change in the current balance $\langle \Delta i \rangle$ in the amount

$$\langle \Delta i \rangle = \frac{\langle \Delta V \rangle}{2L_s} \left(\frac{1}{f} \right) \tag{9.36}$$

This change in current is to be balanced by varying the delay angle. The value Δi_L described in equation (9.35) occurs at every switching pulse. Therefore, the change in a full cycle from variation of delay angle is six times the value of equation (9.35), i.e.,

$$\left[\frac{3V_m}{2\omega L_s} \Delta\alpha \sin\alpha \right] = \frac{\langle \Delta V \rangle}{2L_s} \left(\frac{1}{f} \right),$$

$$\frac{\langle \Delta V \rangle}{V_m} = \frac{9}{\pi} \Delta\alpha \sin\alpha. \tag{9.37}$$

For example, if the average imbalance between bridges is about 3%, then $\Delta\alpha$ must be about 1° when α is 30° if the currents in the two bridges are to be kept balanced.

An oscillograph of the current balancing process during a cycle is shown in Figure 9.10. The voltage across the choke is shown on the upper half of the oscillograph, and the current in the choke is shown on the lower half. The voltage spikes at every switching time are apparent. The rapid rise of current during each voltage spike is also apparent. In this particular example, the current oscillates up and down because an 8-bit microprocessor was used for control and the phase angle delay is divided into coarse steps leading to overcompensation for every cycle. This control technique is a bang-bang technique with similar characteristics to that described in section 7.1.

Control of multiple bridges can be achieved in the same way as described for two bridges. As already indicated, this is essentially the technique used for multiterminal operation of high-voltage direct-current power transmission systems.

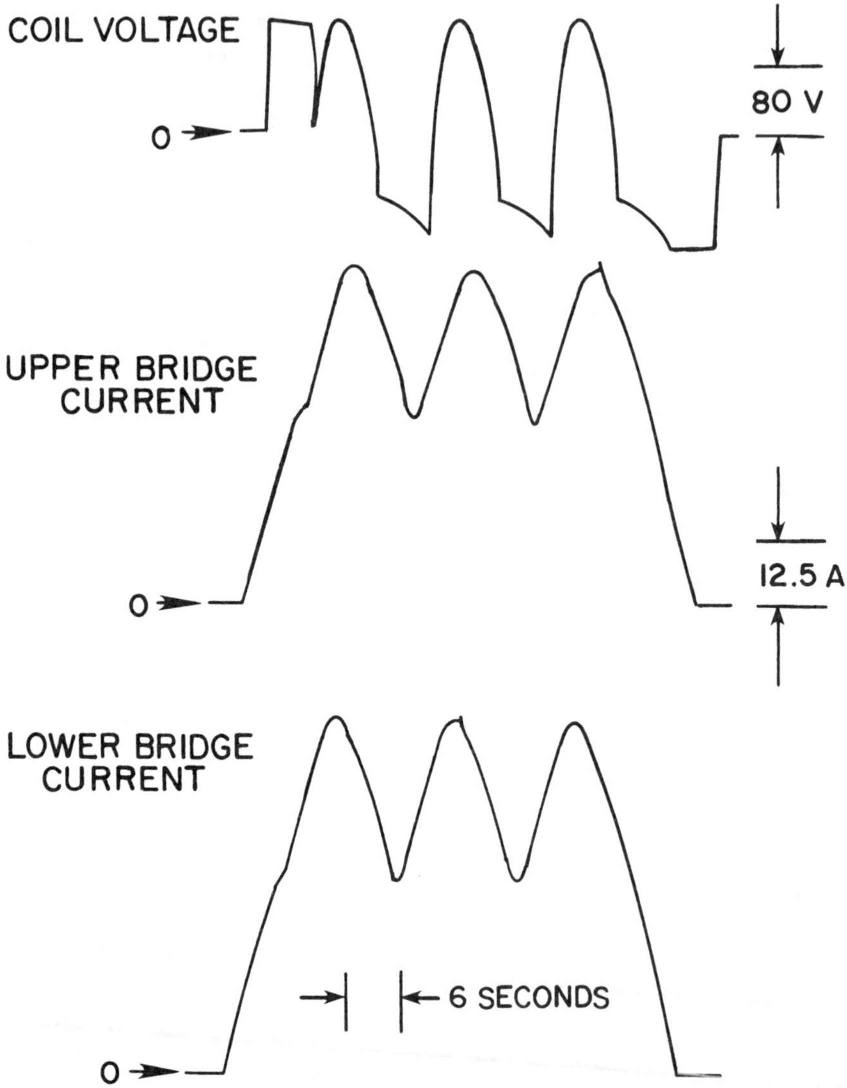

FIG. 9.10. Oscillograph of individual bridge current and loop voltage between two bridges during a cycle when a difference in delay angle exists.

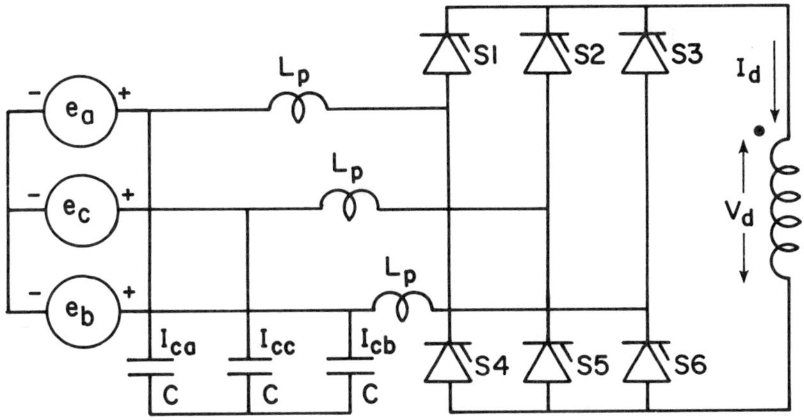

FIG. 9.11. Wye-connected var compensation capacitors.

9.5 Circuit Control Considerations

9.5.1 Var Control with Capacitors on the Utility Side

A standard technique for removing inductive vars is to add sufficient capacitors to provide compensation. A wye-connected capacitor set is shown in Figure 9.11. The following conditions must be met to compensate for the inductive vars:

$$Q = \frac{3}{2}\left(\frac{V_m^2}{X_c}\right), \tag{9.38}$$

or the value of C must be

$$C = \frac{2}{3}\left(\frac{Q}{\omega V_m^2}\right). \tag{9.39}$$

In order to compensate for the inductive vars over the full diurnal cycle shown in Figure 9.7, it is necessary that C vary accordingly. In reality, any power grid will be able to stand a certain amount of inductive or capacitive vars without risking system instability or becoming excessively inefficient. The designer will simply have several capacitors that can be switched in and out at various threshold values of Q.

A typical scenario for $X_p' = 0.2$ is listed below for the diurnal variation given in Figure 9.7.

Threshold Value of Q (mvars)	Value of C(farads)	Capacitive Energy (megajoules)	Switch-in Time (hrs.)	Switch-out Time (hrs.)
400	0.589	1.06	—	—
600	0.884	1.59	4.6	23.0
800	1.179	2.12	5.2	20.6
1000	1.473	2.65	5.7	17.9
1200	1.768	3.18	6.4	14.4
1400	2.063	3.72	7.8	10.3

The use of switched capacitors will effectively work. The shortcomings of the system are in reliability, maintenance, and flexibility. Power systems engineers generally avoid using mechanical switching systems that are cycled frequently because they require continuous maintenance and do not have a high reliability record.

Another consideration is the great flexibility afforded by the Graetz bridge and SMES coil. The idealized power profile shown in Figure 9.5 does not include the rapid power changes that occur during a typical operating period. In a real system, these variations exist and must be handled by either rotating reserve or peaking units using diesel or gas turbine drivers. Costs per unit power with peaking units are normally several times greater than base-load generation, and likewise rotating reserve comes as major expense to the system. The Graetz bridge and SMES coil can swing from no load to full load in a matter of a few cycles. This capability could be put to use in meeting the short-term peaking demands, but a mechanically switched var control system would increase the maintenance and failure mode problems by sizeable factors. SCR switches have been used in place of mechanical switches for some high-voltage dc installations. Their use is strictly a question of economics.

9.5.2 Var Control with Tap-Changing Transformers

As can be seen in equations (9.28) and (9.30), any reduction in V_m will force α to become closer to 0 or π to maintain a given value

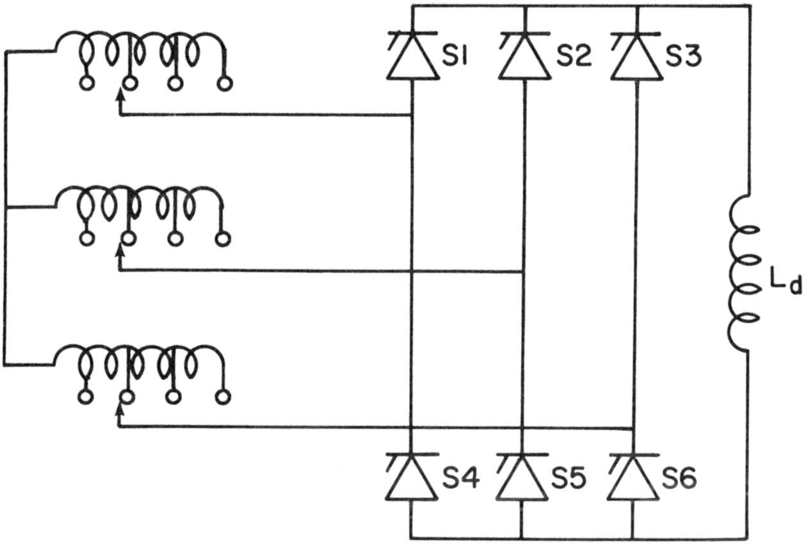

FIG. 9.12. Graetz bridge and SMES with tap-changing transformer.

of P. This directly helps in reducing Q. A tap-changing transformer, shown in Figure 9.12, can be considered for this purpose.

Assume that the circuit employs a transformer with n taps. The transformer will operate at some time in the kth tap and eventually will be switched to the $k + 1$ tap when required voltage on the coil drops sufficiently low.

The line voltage on the utility side of the tap-changing transformer is

$$V_{\ell n}' = 1/3(1 + X_p'/2) \text{ pu}. \tag{9.40}$$

At a current of I_d', the VA value without tapping down is

$$VA = 3V_{\ell n}'I_d',$$

$$VA = (1 + X_p'/2)I_d'. \tag{9.41}$$

When the transformer is on the kth tap, the voltage on the Graetz bridge is

$$V_{\ell n}'(k) = 1/3\left(1 - \frac{k}{n}\right)(1 + X_p'/2)\text{pu}. \tag{9.42}$$

The *VA* value on the kth tap is

$$VA(k) = \left(1 - \frac{k}{n}\right)(1 + X_p'/2)I_d'. \tag{9.43}$$

The power into the coil at the point where the kth tap is first switched is

$$P = \left(1 - \frac{k}{n}\right)I_d'. \tag{9.44}$$

The difference between $VA(k)$ in equation (9.43) and P in equation (9.44) is a result of the overlap angle, which reduces the average voltage on the coil in the amount $1/(1 + X_p'/2)$. Just prior to stepping down the voltage by one more setting to $k+1$, the demand on the average coil has to have been reduced to $[1 - (k+1)/n]$pu; otherwise, the power demand could not be met. Therefore, after the tap change, the line voltage has been reduced to

$$V_{\ell n}' = 1/3\left(1 - \frac{k+1}{n}\right)(1 + X_p'/2). \tag{9.45}$$

The *VA* value after the tap change is reduced to

$$VA = \left(1 - \frac{k+1}{n}\right)(1 + X_p'/2)I_d', \tag{9.46}$$

and the power is

$$P = \left(1 - \frac{k+1}{n}\right)I_d'. \tag{9.47}$$

The values of Q can be calculated at the points where the kth tap was first engaged, just prior to the change to $k+1$ and just after the change to $k+1$. Just after the switch to the kth tap, the power is given by equation (9.44) and *VA* is given by equation (9.43). Therefore,

$$Q = \sqrt{\left[\left(1 - \frac{k}{n}\right)(1 + X_p'/2)I_d'\right]^2 - \left[\left(1 - \frac{k}{n}\right)I_d'\right]^2},$$

$$\simeq I_d' \sqrt{\left(1 - \frac{2k}{n}\right)(1 + X_p') - \left(1 - \frac{2k}{n}\right)},$$

$$\simeq I_d' \sqrt{\left(1 - \frac{2k}{n}\right)X_p'}. \tag{9.48}$$

Just prior to the $k + 1$ tap change, the *VA* value is still given by equation (9.43), but the power has been reduced to the value given by equation (9.47). Therefore,

$$Q = \sqrt{\left[\left(1 - \frac{k}{n}\right)\left(1 + \frac{X_p'}{2}\right)I_d'\right]^2 - \left[\left(1 - \frac{k+1}{n}\right)I_d'\right]^2},$$

$$\simeq I_d' \sqrt{\left(1 - \frac{2k}{n}\right)(1 + X_p') - \left(1 - \frac{2(k+1)}{n}\right)},$$

$$\simeq I_d' \sqrt{\frac{2}{n} + \left(1 - \frac{2k}{n}\right)X_p'}. \tag{9.49}$$

Just after the switch to the $k + 1$ tap, the *VA* value drops to that given by equation (9.46). Therefore,

$$Q = \sqrt{\left[\left(1 - \frac{k+1}{n}\right)(1 + X_p'/2)I_d'\right]^2 - \left[\left(1 - \frac{k+1}{n}\right)I_d'\right]^2},$$

$$\simeq I_d' \sqrt{\left(1 - \frac{2(k+1)}{n}\right)(1 + X_p') - \left(1 - \frac{2(k+1)}{n}\right)}$$

$$\simeq I_d' \sqrt{\left(1 - \frac{2(k+1)}{n}\right)X_p'}. \tag{9.50}$$

Basically, the tap-changing effect can be visualized for a constant

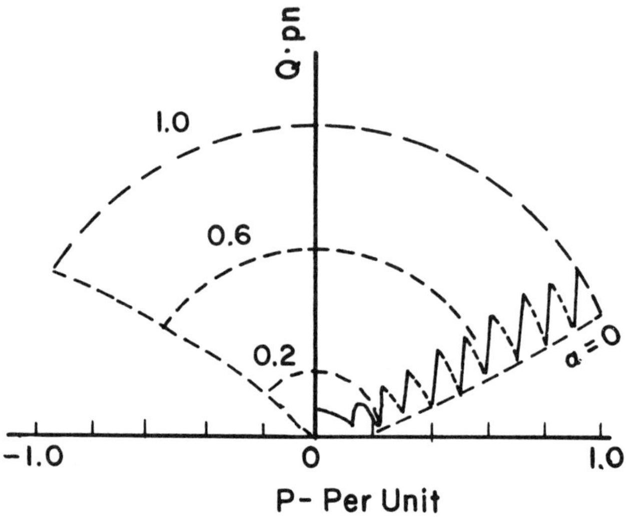

FIG. 9.13. P-Q variation with ten full-range tap changes.

I_d' in the P-Q diagram of Figure 9.13. As available power from the grid drops, the coil voltage needed to match the power requirement also drops. Before the tap change, α is varied from 0 to whatever value is needed to generate the correct average voltage. However, a tap change becomes possible when the required voltage drops below $(1 - 1/n)$pu. The effect of the tap change is to bring the converter delay angle back towards $\alpha = 0$, and P-Q locus now follows the *VA* circle defined by the new voltage. The process is repeated from one tap to the other.

The problems of maintenance and reliability with tap-changing transformers are the same as those for switched capacitors.

9.5.3 Var Control Using Modular Construction of the Graetz Bridge

The Graetz bridge can be divided into modules as shown in Figure 9.14. With this system, each module delivers 0.5 pu of coil voltage. It is possible to operate the two bridges with different delay angles, α' and α'', respectively, provided that the sum, $\cos \alpha' + \cos \alpha''$, can be made equal to the value required for generating the correct coil voltage. In the absence of phase reactance, the power delivered to the load is

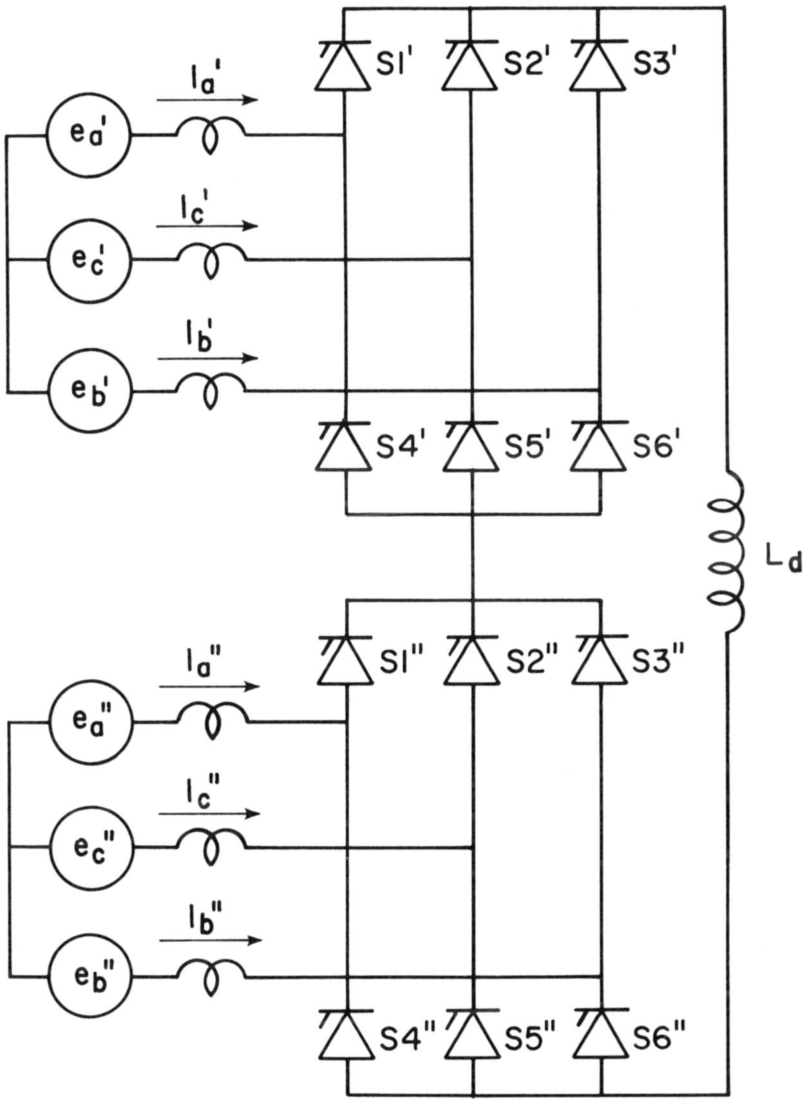

FIG. 9.14. Modular circuitry for Graetz bridge connection to an SMES coil.

$$P = 0.5 \cos \alpha' + 0.5 \cos \alpha'', \tag{9.51}$$

and the reactive power generated on the line is

$$Q = 0.5 \sin \alpha' + 0.5 \sin \alpha''. \tag{9.52}$$

The relative effect of varying α' and α'' for two different power levels is shown in Figures 9.15 and 9.16. For each case, the value of the sum, $\cos \alpha' + \cos \alpha''$, is fixed at the desired power level, and the graph shows the value of α'' needed to maintain $\cos \alpha' + \cos \alpha''$ as a constant value and the variation of Q as α' varies. It is apparent from

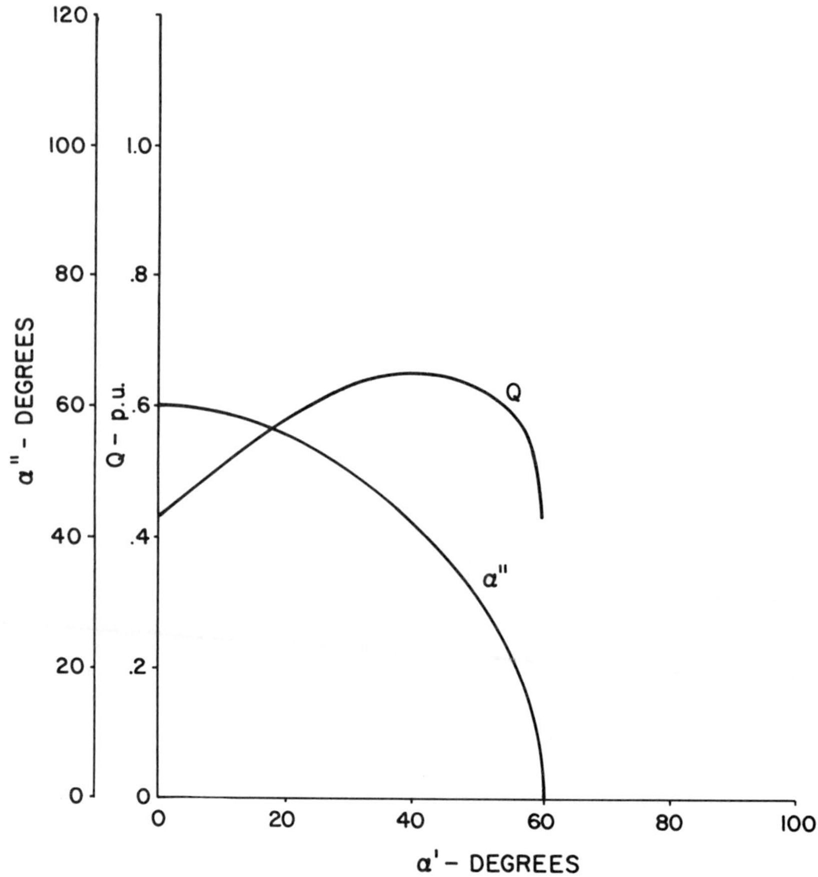

FIG. 9.15. Plot of Q and α'' vs. α' when $p = 0.5 \cos \alpha' + 0.5 \cos \alpha'' = 0.75$.

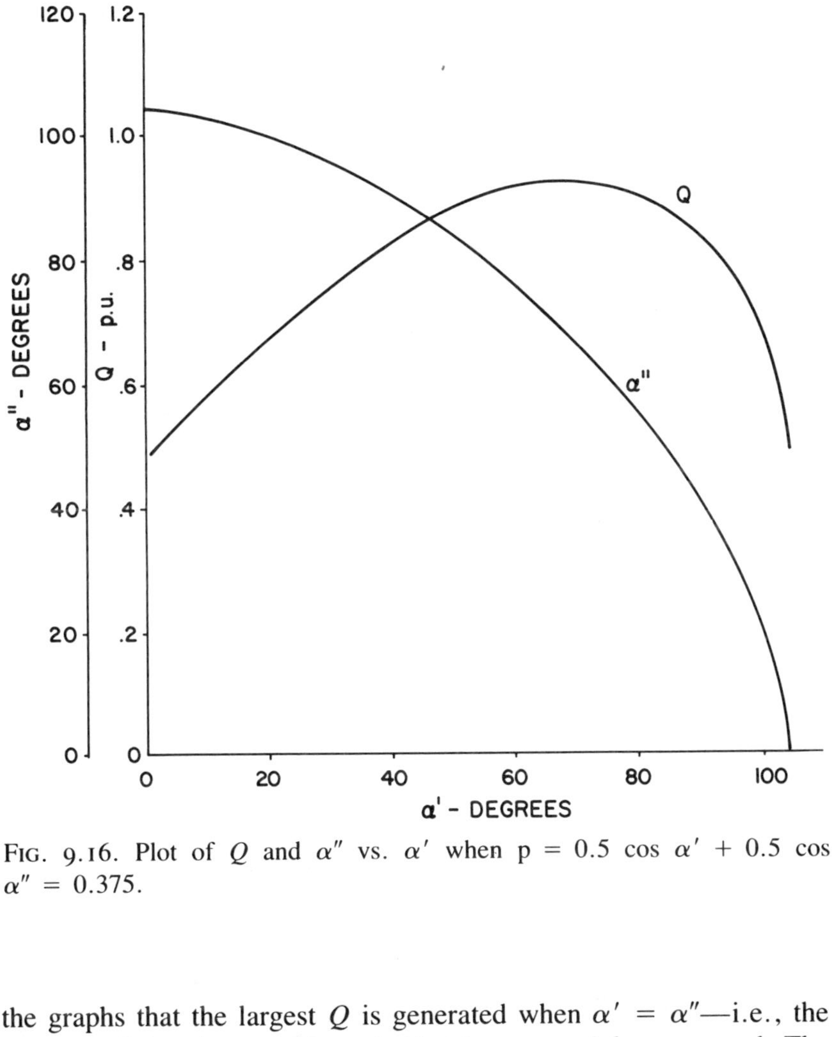

FIG. 9.16. Plot of Q and α'' vs. α' when p = 0.5 cos α' + 0.5 cos α'' = 0.375.

the graphs that the largest Q is generated when $\alpha' = \alpha''$—i.e., the same condition that would result if only one module were used. The lowest value of Q occurs when $\alpha' = 0$ or $\alpha'' = 0$.

The system could be built with more than two modules. With three modules, the per-unit power of 0.375, for instance, could be achieved by operating two of the three modules at $\alpha = 0$, making them contribute a total power of 0.667 pu. The last module must be operated at 151° to provide a negative power of -0.292 pu. The more modules a system has, the more it is possible to minimize Q over the full range of operation. The operating philosophy is to run as many modules as possible with phase angles of 0 or π in order to match the final power with the most modules at a zero Q point.

If one considers Figure 9.14, it becomes apparent that what is done with the modules is to add the current elements back onto the utility with various phase angles generating by the delay angles α', α'' . . . , etc. In vector notation, the utility voltage can be taken as reference, and the vector currents can be added together to get the P and Q values. This is demonstrated in Figures 9.17(a) and (b) for three-module and four-module systems, respectively.

In Figure 9.17(a), the vectors i' and i'' add directly since $\alpha = 0$

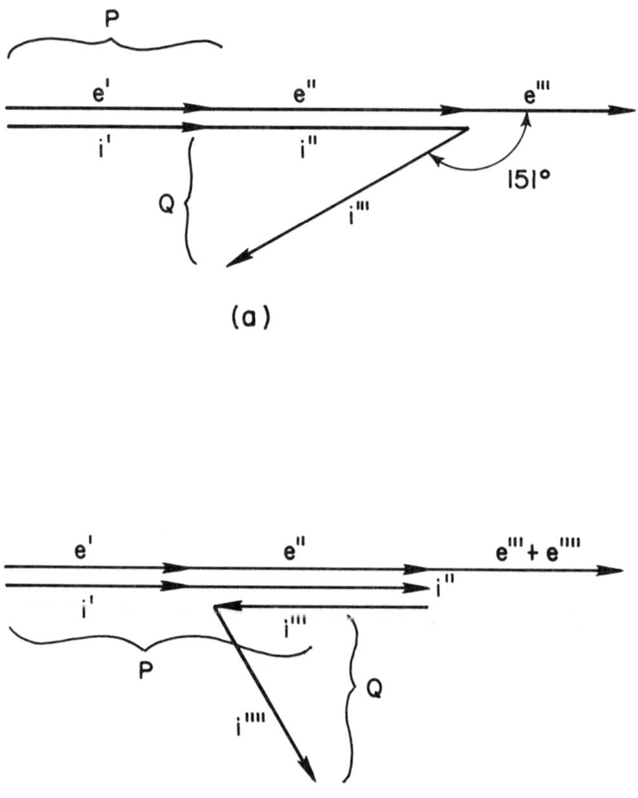

FIG. 9.17. (a) Three-module vector construction for $P = 0.375$ pu; (b) four-module vector construction for $P = 0.375$ pu.

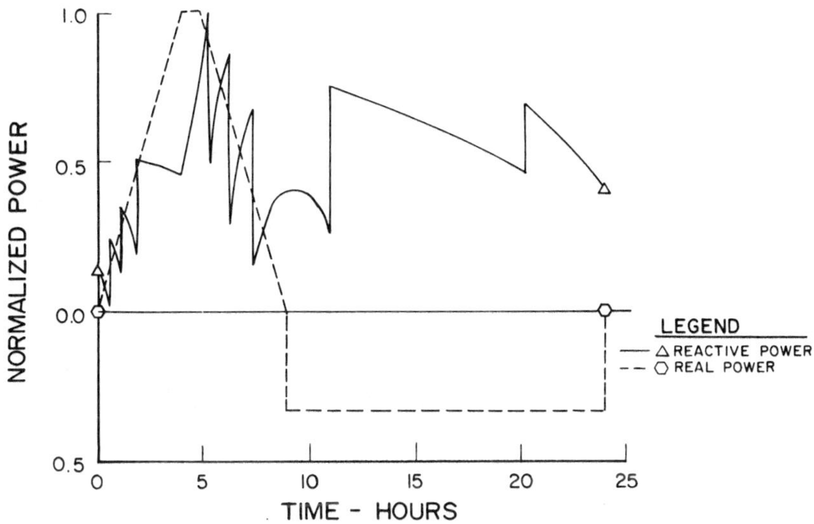

FIG. 9.18. Q values for modular converter during a typical charging cycle.

for these two modules. The third module is operated at $\alpha = 151°$ to provide the correct value of real current to match the required power value, and it leads to whatever value of Q that results to get the three equal-magnitude current vectors to the correct real current.

The four-module system shown in Figure 9.17(b) produces a higher value of Q than a three-module system for $P = 0.375$. This is a direct result of the constant vector magnitudes. At any given particular point, a worse Q condition could be generated with more modules. However, a system with more modules provides greater flexibility and will produce smaller Q values over the full range of power than a system with fewer modules.

The effect of X_p' will cause a phase shift in the currents by the amount roughly equal to the overlap angle. The net result will be to follow constant *VA* circles on the *P-Q* diagram shown in Figure 9.6. A comparison between one-, two-, three-, and four-module systems is shown in Figure 9.18 for the example given in Figure 9.7.

9.5.4 Asymmetrical Gating and Bypassing Methods

As indicated by equation 9.30, the volt amperes under ordinary circumstances are fixed by the line voltage and the coil current. In section 9.5.3, the line voltage was changed using a transformer. With asymmetrical gating the line voltage remains fixed, but the effective line current is changed by shorting, or bypassing, the coil terminals for a part of the ac cycle so that the width of the current pulse that is carried on the ac line is narrower than the full 120° of positive and 120° of negative conduction time. Thus, the peak and rms values of the fundamental harmonic decrease, even though the peak value of the current pulse in the ac line is still equal to the coil current.

Asymmetrical gating is achieved by gating the three SCRs on the top of the bridge S1, S2, and S3 at a different phase angle than the three SCRs S4, S5, and S6 on the bottom of the bridge in Figure 9.2. For parts of the cycle S1 and S4, S2 and S5, and S3 and S6 are simultaneously conducting, producing the short-circuit across the coil terminals. The line voltage is not shorted during this period because, for instance, when S1 and S4 are conducting, S2, S3, S5, and S6 are not, leaving the other two phases disconnected from the coil.

Commutation can take place on the SCRs in Figure 9.2 whenever the line-to-line voltages drive the anodes positive. For instance, the commutation of S3 to S1 requires $V_a - V_b > 0$. For the line-to-neutral voltages given in equation 9.1, this can occur when

$$V_a - V_b = \sqrt{3}\, V_m \sin(\omega t - 30°) > 0. \tag{9.53}$$

For the SCR S1, ωt must be greater than 30° and less than 210° for commutation to occur. The delay angle is defined from 0° to 180°. The upper half delay angle, α', corresponds to 0° when gating occurs at $\omega t = 30°$ and to 180° when at $\omega t = 210°$.

For the SCR S4, the requirement for commutation to take place requires

$$V_b - V_a = -\sqrt{3}\, V_m \sin(\omega t - 30°) > 0. \tag{9.54}$$

For S4, ωt must be greater than 210° and less than 390°. The lower half delay angle, α'', corresponds to 0° when gating occurs at $\omega t = 210°$ and 180° when at $\omega t = 390°$.

Ordinary symmetrical gating occurs when $\alpha' = \alpha''$ and the gating pulses for S1 and S4 occur 180° apart in the cycle. If, for example, α' is chosen to be 180° and α'' is chosen to be 0°, then both of the gating pulses to S1 and S4 would occur simultaneously at $\omega t = 210°$, and the conduction periods of S1 and S4 would fully overlap. This action would bypass the coil current for the full cycle as the conducting pairs switch from S1 and S4 to S2 and S5 and to S3 and S6. The ac line would never be connected across the coil, and no power would be transfered between the coil and the utility. No current would flow in the phases of the ac lines, and no reactive power would be generated either.

Denote the delay angle for the SCRs S1, S2, and S3 by α' and for the SCRs S4, S5, and S6 by α''. In Figure 9.19(a), the line-to-

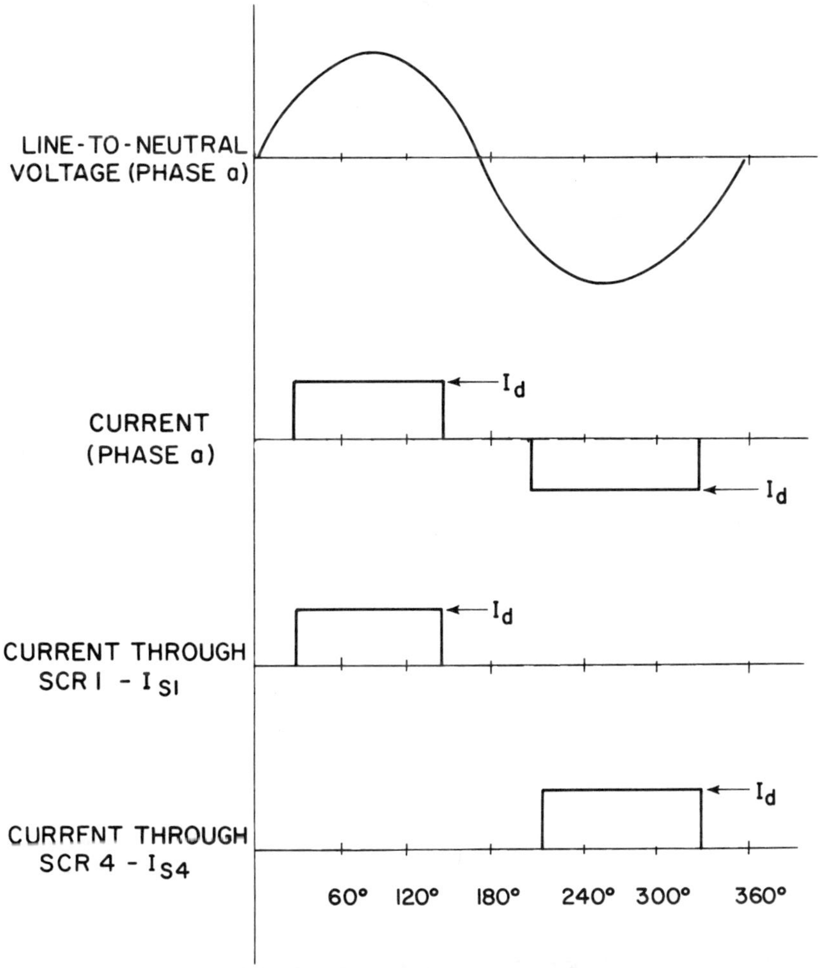

FIG. 9.19(a). Line-to-neutral voltage and current for phase a.

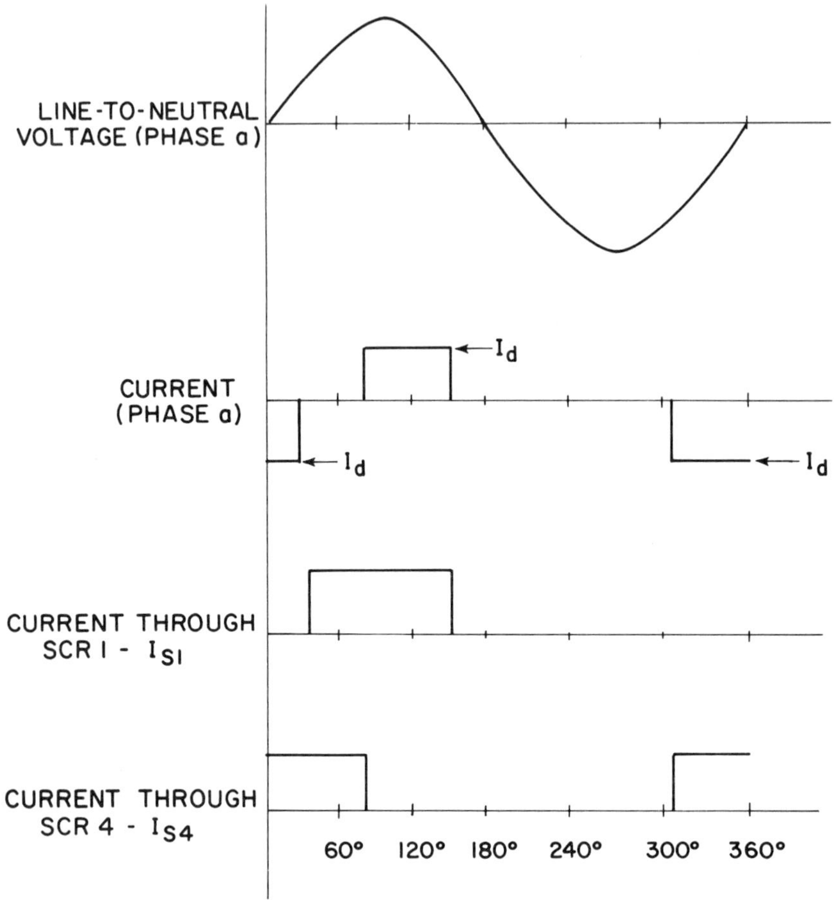

FIG. 9.19(b). Line-to-neutral voltage and current for phase a with asymmetrical gating.

neutral voltage and current for phase a are shown with $\alpha' = \alpha'' = 0$. The current through S1 is shown as I_{s1} and through S4 as I_{s4}. The same waveforms with asymmetrical gating are shown in Figure 9.19(b). The current in phase a is still the same amplitude, but the pulse width is shortened by the amount of overlap. The fundamental component of the current decreases as the two angles, α' and α'', become increasingly different. The real and reactive power normalized to unity are:

$$P = (\cos \alpha' + \cos \alpha'')/2 \tag{9.55}$$

and

$$Q = (\sin \alpha' + \sin \alpha'')/2. \tag{9.56}$$

Given a value of P less than 1, the value of Q reaches a maximum when the two angles are equal, and the lowest value of Q is reached when one of the two delay angles is chosen to be equal to zero. For instance, at $P = 1/2$, a choice of $\alpha' = 0$ and $\alpha'' = \pi/2$ gives the correct power, and $Q = 1/2$. Without asymmetrical gating, $\alpha' = \alpha'' = \pi/3$, the power would be 1/2 and $Q = \sqrt{(3)}/2$. The symmetrical gating technique helps reduce the reactive power considerably, but the technique also increases the generation of harmonics on the power line. The phase current in Figure 9.19(b) can be represented by

$$I = \sum_{n=1}^{\infty} a_n \sin n\omega t + \sum_{n=1}^{\infty} b_n \cos n\omega t,$$

where

$$a_n = \frac{I_d}{n\pi} \left\{ [1-(-1)^n]\,(1 + \cos n\alpha'')\cos \frac{n\pi}{6} + [1 + (-1)^n] \sin n\alpha'' \sin \frac{n\pi}{6} \right\}$$

and

$$b_n = \frac{I_d}{n\pi} \left\{ [1 + (-1)^n]\,(\cos n\alpha'' - 1) \sin \frac{n\pi}{6} - [1 - (-1)^n] \sin n\alpha'' \cos \frac{n\pi}{6} \right\}. \tag{9.57}$$

With symmetrical gating, even harmonics are not generated, and neither are the third harmonics. With asymmetrical gating, harmonics as low as the second are generated. This might be objectionable to some utilities and the second harmonic is a difficult frequency to filter.

References

Boom, R. W. "Review of Diurnal Energy Storage at the University of Wisconsin." In *Proceedings of the International Symposium on Superconductive Energy Storage*. Osaka, Japan, October 8–10, 1979.

———, and H. A. Peterson. "Superconductive Energy Storage for Power Systems." *IEEE Transactions on Magnetics,* MAG-8, no. 3 (September 1972).

Forest, R.; G. Heyner; K. Kanglesser; and H. Waldmann. "Multi-terminal Operation of HVDC Converter Stations." *IEEE Transactions on Power Apparatus and Systems,* PAS-94, no. 4 (July/August 1975).

Kimbark, E. W. *Direct Current Transmission.* New York: John Wiley and Sons, 1971.

McMurray, W. "A Study of Asymmetrical Gating for Phase-Controlled Rectifiers." *IEEE Transactions on Industry Applications,* IA-8, no. 3 (May/June 1972).

Peterson, H. A.; N. Mohan; and R. W. Boom. "Superconductive Energy Storage Inductor-Converter Units for Power Systems." *IEEE Transactions on Power Apparatus and Systems,* PAS-94, no. 4 (July/August 1975).

Index